Studies and surveys in comparative education

Science and technology in the primary school of tomorrow

Graham Orpwood and Ingvar Werdelin

A study prepared for the
International Bureau
of Education

Ministry of Education, Ontario
Information Centre, 13th Floor,
Mowat Block, Queen's Park,
Toronto, Ont.   M7A 1L2

unesco

Titles in the series 'Studies and surveys in comparative education':

Carton, M. *Education and the world of work.* 1984.

Fernig, L.R. *The place of information in educational development.* 1980.

Fitouri, C., et al. *Educational goals.* 1981.

Goble, N.M.; Porter, J.F. *The changing role of the teacher: international perspectives.* 1977.

Haag, D. *The right to education - what kind of management?* 1982.

Havelock, R.G.; Huberman, A.M. *Solving educational problems: the theory and reality of innovation in developing countries.* 1977.

Hummel, C. *Education today for the world of tomorrow.* 1977.

Orpwood, G.; Werdelin, I. *Science and technology in the primary school of tomorrow.* 1987.

Published by the
United Nations Educational,
Scientific and Cultural Organization,
7 Place de Fontenoy, 75700 Paris,
France

Printed in Switzerland

ISBN: 92-3-102502-3

© Unesco 1987

Contents

	Introduction *p. 7*
CHAPTER I	The science-technology-education partnership *p. 9*
	Overview of the study p. 13
	Outline of the study p. 15
CHAPTER II	Contexts of science and technology education in the primary school *p. 19*
	The concept of primary education p. 19
	The performance of primary education p. 24
CHAPTER III	Goals and objectives of primary education *p. 49*
	A model of educational goals and objectives p. 49
	Goals and objectives of primary education as a whole p. 50
	Goals and objectives of science and technology in primary education p. 76
	A summing up p. 94
CHAPTER IV	Structure and content of primary school curricula *p. 96*
	Trends in curriculum development p. 96
	The place of science and technology in primary school curricula p. 102
	The content of science and technology education p. 108
	Conclusions about the existing situation p. 114
CHAPTER V	The delivery system in primary school *p. 117*
	The place of the delivery system in the educational process p. 117
	Teaching methods p. 119
	Educational administration and educational organization p. 129
	Evaluation of pupil achievement p. 139

Teacher training p. 142

Consequences for science and technology education p. 145

CHAPTER VI Renewal of science and technology education *p. 148*

The goals and objectives for renewal p. 149

Means for the renewal of primary education p. 156

Effects of renewal of science and technology education p. 167

Frames for the renewal of science and technology education p. 170

Educational renewal and educational problems: a summing up p. 175

CHAPTER VII Reflections and questions for deliberation *p. 177*

Contexts of science and technology education in the primary school p. 177

Goals and objectives of primary education p. 181

The dilemma of diversity: the school's dilemma p. 183

The dilemma of diversity: the teacher's dilemma p. 185

Structure and content of primary school curricula p. 186

The delivery system in primary school p. 190

Renewal of science and technology education p. 195

References *p. 199*

Introduction

The work presented here is the result of a partnership between its two authors and the International Bureau of Education in Geneva. At its twenty-first session, the General Conference of Unesco decided that the special theme of the thirty-ninth session of the International Conference on Education, to be organized by the International Bureau of Education, should be 'Universalization and renewal of primary education in the perspective of an appropriate introduction to science and technology'.

The session was convened by the Director-General of Unesco at the International Conference Centre, Geneva, from 16 to 25 October 1984. It was attended by 123 Member States of Unesco and several other organizations, both governmental and non-governmental. In all 466 delegates, including 34 ministers and 24 vice-ministers, and 74 other representatives and observers were in attendance.

The *Final report* [34][1] of the Conference gives an account of the conclusions of the debate of Commission II which discussed the special theme and which prepared a formal recommendation to ministers of education. This recommendation was subsequently discussed and adopted by the plenary session of the Conference.

The special theme of the Conference points to a need for renewal of primary education to provide for universal schooling in countries where it has not yet been attained. Science technology education is seen as essential to this renewal. The problem expressed by the theme is of general concern to education authorities in the Member States in all parts of the world. Despite the huge efforts that have been made to advance education and despite the progress achieved, the full exercise of this right is still far from being world-wide. The choice of the theme of universalization and renewal of primary education reflects the growing preoccupation of education authorities in many Member States with the alarming situation concerning illiteracy, which to a certain extent thrives on the inadequacy of primary education. In addition to the problem of general provision for and further progress towards the democratization of education, the content and methods of primary education are often not adapted to the natural, cultural and human environments in which this education is provided; the content and curricula often lag far behind the advancement of knowledge, particularly as regards science and technology education in primary school.

The problem of renewal to achieve generalized primary education is also stressed by Unesco. In its second Medium-Term Plan (1984-1989) it states:

...if the democratization of education implies the expansion of education systems to enable the schools to cater for the whole school-age population, it also involves in many countries an

---

1. The numbers in square brackets refer to the 'References' at the end of the book.

effort to renew the organizational framework, methods and subject-matter so as to ensure that education is more relevant to the environment in which it is given, the needs of the people and the requirements of progress.

Recent conferences of Ministers of Education and Those Responsible for Economic Planning in Member States in the different regions of the world, particularly the conferences held in Mexico City for the Latin American and Carribean Region [63] and in Harare for the African Region [6], expressed concern for the problems of the universalization of primary education and recommended steps for its further democratization and renovation in order to provide all children with education such that they can exploit their own aptitudes and enable society to progress. Similar concerns were expressed at some of the conferences regarding the development and renovation of science and technology education at all levels.

To obtain up-to-date information on which to base the discussion during the Conference and to enable the writing of the present book, the International Bureau of Education conducted an inquiry by sending a questionnaire in 1983 to all Member States of Unesco. Before the deadline, eighty-one countries had provided answers to the questionnaire. They were asked to provide information on several points concerning: (a) universalization of primary education; (b) renewal of primary education; (c) introduction to science and technology at the primary level; and (d) regional and international co-operation

During the conference other official documents from Member States were made available. Furthermore, a large number of international and national meetings have dealt with the topics of the Conference and have led to the publication of documents, and much research has dealt with this subject.

Following the Conference session, the International Bureau of Education commissioned one of its comparative studies on the same general theme as that of the Conference, and this book is the outcome of that study. The authors attended the Conference; one of them (Werdelin) produced a paper for it. The planning of the study was a joint effort. One of the authors (Werdelin) was responsible for reviewing the information submitted by the Member States and other materials and for preparing Chapters II to VI inclusive of this book. The other (Orpwood) wrote Chapter I, which describes the conceptual framework of the study, and conducted a reflective analysis of data, presented in Chapter VII. Despite the fact that at no time during the writing phase of the study did the authors live closer to each other than 8,000 kilometres and that they only met for one week to discuss the writing, the collaboration was a stimulating and happy one. Whether it was also productive, the reader alone must judge.

Each of us wishes to acknowledge institutions that employed him. Werdelin prepared much of the text while on study leave at the School of Education, University of Melbourne, Australia, and expresses his gratitude to that institution as well as to his home base, the Department of Education and Psychology, University of Linköping, Sweden. Orpwood received encouragement and co-operation from the Science Council of Canada, despite the fact that it was undergoing a severe reduction in its resources at the time. To each of these institutions, and especially to our personal colleagues, we say 'Thank you'.

CHAPTER I
The science-technology-education partnership

The citizens, workers and decision makers of the twenty-first century are in school today. Are they receiving the education they will need to enjoy full, happy and productive lives? This question is of pressing concern to education authorities throughout the world as they try to develop policies and curricula appropriate for their societies, which are changing at an ever-increasing rate. One of the principal forces driving these changes is science and technology; education in science and technology is, therefore, under particular scrutiny in many countries of the world at this time.

This interest in science and technology education also implies a continuing belief, on the part of all nations, in the potential of education to contribute to the betterment of their peoples. Science and technology, despite their awesome power, can ultimately be controlled and used for peaceful, socially useful ends, if they are in the hands of truly educated people — men and women whose moral and intellectual capacities have been developed to the full. An introduction to science and technology at the primary level of education is thus seen as essential to this goal.

We can illustrate this ideal partnership of education, science and technology in support of national development in the form of a simple diagram (see Figure 1).

It is the premise of this book that, if economic and social development are to progress smoothly and effectively, these three contributing enterprises must function as partners. Historically, this has rarely been the case (even in the most developed countries of

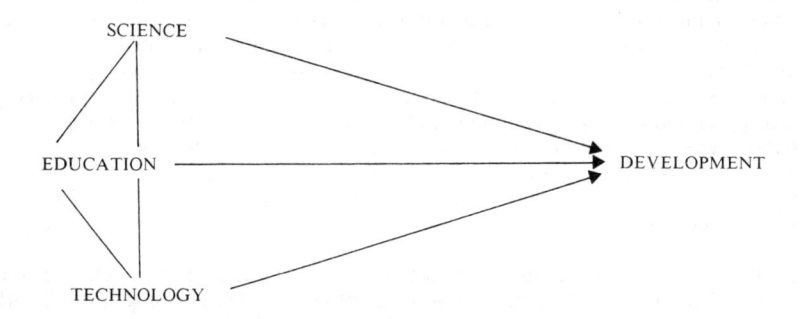

FIGURE 1. Science, technology and education: a partnership for development

the world) and thus the quality of science and technology education has continued to be a matter of concern. Is our faith in such a partnership therefore misplaced? *Can* science, technology and education work together for development? Let us examine these questions and at the same time suggest working definitions for the key terms.

Technology

It is common to find definitions of 'science' and 'technology' set down in this order, with 'science' first explained and then 'technology' defined as 'the applications of science' [37, p. ii]. While such a definition of technology is useful for some purposes, it can also be misleading since it casts technology in a derivative role and implies the scientific knowledge must first exist before it can be applied by the technologist or engineer. Such a view of science and technology turns the history of social and economic development on its head since technologies, by any reasonably broad definition, have — until the eighteenth century, at least — usually preceded the corresponding areas of science. It seems appropriate, therefore, in the present context to consider technology first.

As part of their cultural heritage, peoples in all parts of the world have developed tools, materials and techniques to meet their needs and desires. As the raw materials available to them and their own ingenuity have varied, so also do the products of their 'technological' innovation vary in sophistication and endurance. But the basic needs of shelter, clothing, nourishment and physical survival, as well as the human desires for comfort, transportation, communication and military superiority, led to the development of technologies in many areas. This process of technological development is intimately bound to the process of education since improvements to each technology have been linked to their being passed on from father to son and from mother to daughter over many generations. This system of technological education has been one of initiation or apprenticeship and is quite distinct from (though sometimes parallel to) those processes of education carried out formally in schools (referred to here simply as 'schooling'). As each generation learned the traditional crafts from the preceding one, so they added their own improvements and refinements which resulted in greater efficiency or satisfaction.

It is important to realize that this technological education was independent both from science (as a formalized body of knowledge) and from schooling. For example, the working of metals and the making of ceramic pots both preceded (by thousands of years) the scientific discovery of the structure of these materials.

With the advent of science, however, technology has become dependent not just on material resources and human ingenuity but also increasingly on scientific knowledge. Correspondingly, its rate of development has increased enormously and it has become fashionable to talk of technology as 'applied science' since the scientific component has become so important. To continue the example of materials technology, the science of metallurgy has led to the development of many new alloys with special properties and, even more recently, scientific research into ceramics has shown the enormous potential of these materials in future industrial design. Yet the essence of technology remains the same: in the words of our preferred definition, 'the totality of the means employed to provide objects necessary for human sustenance and comfort' *(Webster's new collegiate dictionary,* 1977).

With the advent of science education (see next section), the teaching of craft technologies has taken a minor role in modern education [35, p. 2]. The teaching of woodwork, metal work, needlework and cooking are still found in schools in many countries, but often they are provided only to students who are not considered to have the ability to handle the (more prestigious) academic subjects.

To summarize, technology continues to be of direct importance to the economic and social development of all countries. Nowadays, scientific knowledge is an essential component of technology which is becoming increasingly 'knowledge-intensive'. However, it is our hypothesis that this partnership between science and technology has not been successfully transferred to the classroom in most countries (for an interesting account of the political problems in attempting this transition in one country — the United Kingdom — see [43]). In addition, the partnership between technology and education which existed in traditional cultures has, we hypothesize, not survived the take-over of education by schooling. Before exploring these questions, however, let us consider the second component of the ideal partnership for development — science.

Science

Unlike technology, with its practical goals, science is the search for knowledge, for understanding, for explanations of phenomena. While technology uses concrete materials to build its products, the stuff of science is ideas, the intellectual creations of human beings. At its most basic level, it seeks to describe the objects of nature. In doing so, it invents concepts which function as ways to examine natural phenomena. At the next level, it tries to link these concepts together in ways that can be used to classify objects and events into groups. This is important because individual instances do not then have to be considered as unique, each requiring a separate explanation. For example, once the category 'fish' is distinguished from the category 'bird', it is not necessary to explain why each separate fish swims and each separate bird flies. Science is thus hierarchical in nature, moving toward ever more sophisticated abstractions, ever more minute levels of details, ever more general explanations for the world we live in.

These explanatory systems are not themselves useful in any practical sense. The ancient Greeks developed remarkably advanced scientific ideas but were less sophisticated technologically than either the Egyptians or the Chinese. It was not until the eighteenth century that the potential of science for contributing to technology began to be exploited systematically and it is only in our present generation that the length of time between basic scientific discoveries (e.g. of the biochemical basis of genetic reproduction) and their technological exploitation (in biotechnology, to continue the same example) has become very much shorter.

Along with this shortening in the time between discovery and application has come the tendency mentioned earlier to regard technology simply as 'applied science', thus making the connection between science and technology a sort of 'one-way street' in which scientific knowledge is seen as the prerequisite to technological development. As we have noted, while this is increasingly becoming the case with the newer 'high' technologies, it is still not the case with all technological development. Nor does it do justice to the history of technology in which most technological innovation has been based not on science but on the need to overcome the shortcomings of earlier technologies and on the use of either new materials or technical ingenuity on the part of engineers or technicians.

Of more importance in the present context is the connection between science and scientific activity in a country and the rate of economic development. Partly, we suspect, because of assumptions concerning the direct and linear link between science and technology, a significant research capability in the basic sciences has often been regarded as a prerequisite to technological development and thus to economic success. The United States of America, which clearly possesses all of these, is held as the example that proves the point. Yet it is also clear that countries such as Japan and, more recently, the Republic of Korea have developed phenomenally successful technologically-based industries without first becoming dominant in research in the basic sciences. It is also the case that other countries (such as the United Kingdom) that have long had a tradition of excellence in scientific research have failed to see a corresponding payoff from technology. So, not only is the connection between science and technology complex epistemologically, it is also complex economically, socially and politically. National development requires endogenous technological development but not necessarily a correspondingly sophisticated tradition in the basic sciences.

If this is indeed so, then the implications for education are significant. There are good educational reasons for arguing that science is an important component of a student's general education but it may be that technological education is of even greater importance in stimulating economic development. Ideally, of course, the two will be partners. Let us therefore turn to education and consider how it is expected to contribute to national development.

Education

From the earliest times of civilization in all parts of the world, education, like technology, has had its roots in the home and local culture. Only since knowledge has become largely written down in books has education become associated with literacy and the acquisition of literacy with institutions such as schools. (Schools, that is, in the modern sense of the term. Certain types of school — religious schools for example — are, of course, much older.) In recent times, education has become associated almost exclusively with schools and thus its scope has tended to be limited to the sorts of things that can be readily carried out in schools. For example, schools are quite good places for learning how to read and how to calculate. Reading and arithmetic have thus become important school subjects. Cooking and making tools, however, are less effectively taught in schools than in kitchens and workshops; correspondingly, they have not come to be seen as important aspects of education.

This is not, in any way, to denigrate the value of literacy and numeracy but simply to stress the distinction, introduced earlier, between education and schooling. Education is the more inclusive term and includes those cultural, religious, moral and technical aspects of a child's development that we may choose not to (or be unable to) deal with adequately in schools. The point of the distinction in the present context is that it helps us to understand how education in science and technology has become narrowed as the concept of education has become more and more synonymous with that of schooling.

Unlike technological education, which began at mother's knee and father's field or bench, education in science has always been predominantly a matter for formal schooling. Science became formalized in the school curricula in most European countries during the latter part of the nineteenth century in a variety of forms from health to astronomy and from botany to physics. However, from the earliest days of its introduction, science education was not seen as a subject for all students but as one for an elite who would go on to become scientists, doctors, engineers and leaders of industry. In this respect, it fitted well into the secondary school systems of European countries that were themselves largely elitist in orientation. As these countries developed school systems in their colonial territories, they introduced similarly oriented curricula.

However, in recent years, the advantages of such selective or elitist school systems have been called into question both in Europe and in those contres with school systems based on European models. For example, a recent Unesco Bulletin on Science Education in Asia and the Pacific states:

The consequences (of giving priority to higher levels of education) have been an increase in the number of illiterates, despite the expansion of the formal system; an increase in the number of unemployed or underemployed; and a general support for, in some cases, a small industrial sector based on Western methodology, rather than on the much larger traditional craft economy. Thus the educational system as a whole has continued to support or serve the interests of socially and intellectually privileged groups rather than society in general [37, p. iii].

The outcome of this type of rethinking of the role of schools in relation to national development has been to focus attention more on such matters as literacy and the needs of all society, on the relevance of school subjects to the local culture and economy, and on the connection between education and work. It has also re-emphasized the role of the primary school as being more than simply a place to prepare some students for secondary and higher education.

These trends have been reflected in the way that science education has been critically examined in recent years. In the early years, science in school meant (for most students) studying the separate sciences of physics, chemistry, biology (and sometimes others) as part of a preparation for university or other advanced training. These specialist subjects were part of a secondary school curriculum which, in many countries, was only available to a small proportion of students. Now, however, the idea that science is a subject about which all people should know something requires that it be incorporated into the primary level of education and this trend has been evident worldwide for a number of years. Hence, the growing involvement of international agencies, such as Unesco, in the teaching of primary science. For example, the Unesco series *New trends in biology* (also *chemistry, mathematics, physics*) *teaching* began in the late 1960s, while the first such publication for primary science education was published in 1983 [19].

Part of the motivation for this expansion of science education to include all students and, therefore, into the primary levels of education has undoubtedly been the expectation that the partnership among science, technology and education would be strengthened and that the cause of economic and social development would thereby be furthered. It is beyond the scope of this book to determine if, indeed, economic development has been furthered through changes in education systems. It is probably too early to make such an assessment, in any case. However, we shall examine the ways in which science and technology education have developed at the primary level of schooling and the extent to which science, technology and education are functioning as partners in the service of development throughout the world.

OVERVIEW OF THE STUDY

Background

Evidence on the growing international awareness of the importance and need both to universalize primary education and to introduce science and technology into the curriculum at this level was the incorporation of these two issues into the special theme of the thirty-ninth session of the International Conference on Education held in Geneva, 16-25 October 1984. This theme, as stated for the purposes of the Conference, was 'Universalization and renewal of primary education in the perspective of an appropriate introduction to science and technology'. This theme was discussed by one of the two commissions of the conference, at the end of which a recommendation to Ministries of Education was adopted on the subject. This study is not a formal report on that debate nor on the substance of the recommendation; readers interested in consulting these should refer to the *Final report* [34].

In preparation for the Conference, the International Bureau of Education prepared two questionnaires for Member States. The first of these related specifically to the special theme of the Conference, while the second was devoted to the topic of the plenary session of the Confernce - 'Education for all in the new scientific and technical environment and taking into account disadvantaged groups' [31], and to the theme of the Conference's other commission - 'Major trends in education' [35]. Clearly, there are connections between the plenary theme, which emphasizes the new scientific and technical environment, and the special theme, with its stress on the educational response to that environment through science and technology education at the primary level.

The primary purpose of collecting information through these two questionnaires was the preparation of a database of experiences and opinions in the form of working papers for the Conference. However, these papers were necessarily restricted in scope, length and style for their effective use as conference papers and could not possibly do justice to the wealth of informative data contained in the responses of Member States. In all, responses were received from eighty-one Member States and the present study draws on these extensively (as well as on supplementary information from thirteen other Member States — see 'References—questionnaire replies'). These ninety-four nations represent, in total, the majority of the world's population.

Scope and objectives of the study

The theme of the present study has been foreshadowed in the conceptual overview above. We are concerned to review the degree to which science, technology and primary education are effective partners in the service of national development. The overview set out a series of hypotheses that were not altogether optimistic. The responses to the questionnaires will be used as the basis for an analysis of the quality of the partnership as it presently exists and, more optimistically, of policies and activities designed to strengthen it.

Objectives of the study are twofold: to provide comparative information about the state of science and technology education at the primary level in Member States and to analyse this information in such a way as to raise questions which policymakers and educators must answer within the context of their own situation. It is specifically not our intention to attempt to evaluate national or regional policies or programmes for science and technology education, nor to offer general recommendations for Member States to follow. Not only would it be presumptuous to do so, it would also be unhelpful. Rather, we believe that the best value of a study such as this is to raise questions that might not otherwise be asked and to present data in a perspective that might not otherwise be seen.

For this reason, while our major source of information is the responses of Member States to the IBE questionnaires, it is not our only one. Numbers of publications from Unesco, other international or regional organizations and from national governments or institutions have been made available to us and these are used (and cited) to supplement the IBE data.

However, the authors have not had much opportunity to conduct primary research projects of their own for this study. Some basic research by one of the authors (Werdelin) is presented here for the first time. Most of the remaining data reported here is secondary, in that it has been collected by others. Furthermore, the majority of the information has been supplied by or collected for use by policymakers and this inevitably affects the type of information collected or seen as significant. In a recent national study conducted by one of the authors (Orpwood), it was found useful to understand this phenomenon in terms of four levels in which the science curriculum can be perceived.

At one level, the evidence consists of the program, courses of study or curriculum guidelines prescribed by ministries of education; these define the *intended* curriculum. At a second level, school boards, schools and teachers create the *planned* curriculum through local programs and lesson plans. Thirdly, evidence from the classroom itself indicates the *taught* curriculum, which students actually experience. And finally, there is the *learned* curriculum, the students' intellectual and practical achievements [72, see also 52].

In an ideal world, of course, all four levels would be identical but common sense suggests that this is never the case. This does not imply dishonesty on the part of those reporting evidence but merely the reality of different ways of looking at the phenomena. In the present study, then, the vast majority of our evidence was from the first level.

OUTLINE OF THE STUDY

The study may be considered as comprising three parts. Following this introduction, five major chapters contain the summarized evidence from Member States concerning science and technology education in the primary school. A final chapter contains an analysis of these data using a variety of concepts from the literature of science and technology education theory. The data chapters (II to VI) are based largely on questions from the IBE questionnaires, as follows:

Contexts of science and technology education in primary school

The debates concerning science and technology education at the International Conference on Education took place within the context of a broader discussion of the universalization and renewal of primary education in general. Clearly, the nature and quality of primary education in a country affects directly the nature and quality of any science and technology education that is included. It is appropriate therefore that this study start with a review of Member States' responses to questions concerning their systems of primary education.

Presentation of this information is in two principal sections corresponding to two major aspects of the context in which science and technology education takes place. First is the historical context. Many of the Member States' responses concerning their systems of primary education began with a historical outline of the development of education in their countries. This information is summarized in terms of the principal models of primary education that have been used in the growth of universal primary education throughout the world.

This historical development has led to a variety of concepts of primary education. Indeed, the term 'primary education' varies in meaning around the world. While the IBE questionnaire refers to the International Standard Classification of Education (ISCED) definition of 'education at the first level which usually begins at age 5, 6 or 7 and lasts about five or six years', it also recognizes a variety of more specific national definitions. The responses to the questionnaire enable the study to provide an overview of this variety and to compare the lengths of primary education with those of compulsory education in various systems.

The second aspect of the general context for science and technology education can be described as the performance of primary education. Here we review such indicators as enrolment, wastage (dropping out, repetition, absenteeism, etc.). In connection with the thirty-ninth session of the International Conference on Education, data on educational development in the world were published and trends in these data are reviewed briefly in this study, together with some of the reasons for failure of primary education to achieve its goals. Discrimination in education on the basis of sex, race, geographical location, physical handicap, language and other factors is another component of the overall quality of education, and progress towards reducing it is discussed here.

Goals and objectives of primary education

At the primary level of education, science and technology form part of a total curriculum which is often taught in an integrated manner, i.e. without division into separate subjects. In any event, the goals and objectives of science and technology education in the primary school are derived from, or at least related to, the general goals set for primary education. This chapter is therefore divided into two principal sections, the first of which reviews the goals and objectives of primary education in general, while the second contains a summary of the goals and objectives of the science and technology component of primary education. (This contrasting layout — first describing primary education in general and following with the science and technology component — is employed in each of Chapters III to V.)

The goals and objectives themselves have been analysed using a two-dimensional model. One dimension considers the degree to which the goals are centred on the needs of the individual or those of society, while the second distinguishes between goals that are concerned with change and development (either in the individual or in society) and those that imply a more stable or traditional view of either the individual or society. The latter casts the learner in a more passive or reflective role while the former sees him or her in a more dynamic or active role. No judgements are implicit in this classification. It is our experience that school curricula in most countries contain goals and objectives of all these types. Therefore, this classification enables ready comparison of goals from radically differing cultures and educational traditions.

While the general goals of primary education are often very broad, those for science and technology education are frequently much narrower. The general goals stress the need to develop literacy and to help young people identify with their culture and ideology, but few countries specify the scientific or technological aspects of contemporary culture. They are, in general, still not regarded as very significant components of the environment and specific references are rarely found in the goals of primary education.

Nevertheless, most countries do acknowledge a place for science and technology in the primary education of their children, a fact that would probably not have been evident twenty years ago. The chapter contains a brief review of the development of increasing international agreement over the past two decades concerning the importance of science and technology education; it then goes on to review the objectives of science and technology education using the model that was described earlier.

Structure and content of primary school curricula

Debates about goals and objectives in education inevitably have an air of abstractness, almost of unreality, about them. Certainly in terms of policy statements, we seem to be a long way from the classroom. Yet once the discussion turns to content, to the stuff of science and technology, then that air changes dramatically toward the concrete, even the mundane. Yet content is important in giving flesh to the abstractions of goal statements. It is also important since the content of science lessons provides the 'vehicle' for carrying the objectives. This concept will be explored in more detail in the final chapter as we discuss how the *same* science content can be used to achieve a variety of *different* goals or objectives.

First, the chapter reviews the general structure and content of curricula in countries around the world. Within this section there is a discussion of integration (of school subjects) within the curriculum. Following this, there is a more detailed analysis of Member States' responses to questions concerning the science and technology curricula of their schools. These responses are important to the study inasmuch as they shed light on the hypotheses advanced earlier in this chapter concerning the partnership (or lack of one) among science, technology and education.

The delivery system in primary school

Even the best science or technology content combined with the most useful objectives in a curriculum plan would mean nothing at all if they were not 'delivered' appropriately to the learner. As the earlier classification of curriculum levels implied, there is often significant slippage between the intended and planned curriculum (on the one hand) and the taught curriculum (on the other). Everything that affects this gap is included in what we call the 'delivery system'. Delivery systems therefore include most importantly the teachers (and, by extension, their training); also, textbooks and other instructional aids such as radio, television, computers and other educational technology, where this is used. They also include, in the case of science and technology education, science rooms or laboratories, technology workshops, and the equipment that might be found in such facilities.

These factors have been extensively studied together with their effects on the outcomes of the learning process. We do not attempt in this study to review that research in detail but rather to summarize the responses of Member States to questions about these factors within their own school systems. In addition, this chapter of the study reviews systems of educational administration and the evaluation of student achievement, both of which have a major impact on the way science and technology are taught and learned. Finally, there is a section that reviews the teacher training provided for teachers of science and technology in primary schools.

Renewal of science and technology education

A review of the state of science and technology education in the world at a given point in time is necessarily a kind of still photograph and may, indeed, be somewhat discouraging. Yet the moving image is much more optimistic, since what we see is less the inadequacies and more the trends towards overcoming them. The thirty-ninth session of the International Conference on Education conveyed impressions of both these images and it is therefore appropriate that our data section of the study concludes with some information concerning the processes of renewal of science and technology education, as these are reported by Member States.

Accordingly, Chapter VI contains summaries of information provided in response to questions concerning these renewal processes. Many countries report such trends occurring at the present time and give details of their major features. The scope and extent of the renewal process are also discussed, along with factors affecting them.

Conclusion

The final chapter of the study has two functions. The first and major part of the chapter is to review and analyse the data already provided in Chapters II to VI. This part will take the form of five reflective essays based on the data of the earlier chapters. The purpose is to raise questions for policymakers and other educators to ask themselves within their own situations. These essays will also introduce a variety of concepts from the literature of curriculum theory and of science and technology education.

The second function of the chapter is to revisit the original hypotheses of the study and to ask in more general terms, 'In the light of the information provided (in Chapters II to VI), what can be said of the partnership between science, technology and education?'. Thus the study will conclude, as it began, with a concern for the economic and social development of all countries and of the role and effectiveness of science, technology and primary education to work together to further these goals.

CHAPTER II
Contexts of science and technology education in the primary school

In this and the following three chapters, an attempt will be made to paint a picture of science and technology education in primary school throughout the world. This will mean discussing goals and objectives; describing curricula; mentioning means used in the introduction of science and technology in primary school teaching; and discussing difficulties encountered, ways of overcoming them, and factors favourable to the process. To be able to present an adequate picture, it is first necessary to sketch the background and to see science and technology in relation to the total educational situation.

A first question that needs to be answered is what is meant by 'primary education' in different countries. The term is used with different meanings in different parts of the world, and it might be misleading to discuss science and technology education at this level without having made clear what stage of education is considered.

A second question that must be asked about primary education concerns the ways in which it functions in the world: do different countries get the expected outputs from the economic and other investments made? Different school systems operate at various levels of efficiency, and in some developing countries it is found that inefficiency in the area of education means a hindrance to the whole national effort to improve the socio-economic situation: a large number of children never enter school; the majority or a large proportion of those who enter school never finish primary education; others repeat one or several grades; and discrimination between groups in society is common. It is felt that proper approaches to this educational problem should enable countries to improve their whole situation, and that science and technology should play a major role in this work since these subjects could bring school closer to practical work, local realities and future opportunities. On the other hand, the opportunities for introducing young people to science and technology depend on the ways in which the school system function.

THE CONCEPT OF PRIMARY EDUCATION

Before a study is made of goals and objectives, curricula, delivery systems and other aspects of primary education, it is necessary to realize that primary education systems themselves differ fundamentally from one country to another. They are based in part on different traditions, and they have developed in different directions.

The role of primary education
Looking back to the educational situation existing only a century ago, one finds that the concept of broadly based primary education for everybody had not generally entered

people's minds. In most countries, the education provided for the masses was closely related to their religious needs and had little or nothing to do with preparing them for active life. Examples may be taken from Sweden, where public education was provided by the State Church since the late seventeenth century and aimed to teach people to read the Bible; or from Moslem countries, where mosque schools taught children to recite the Holy Quran in classical Arabic. Schools existed for a social and intellectual elite, too. They provided traditional academic education for children who would pursue careers in religious, legal, medical, administrative and similar fields. Scholarship systems seem to have existed in most countries enabling gifted children from poor homes to enter these elite schools and climb the social ladder.

During the nineteenth century, most countries then existing tried to generalize education at the primary level and usually made it compulsory. Laws about compulsory education often go back a long time; the following years may be mentioned: Austria, 1774; Denmark, 1814; Turkey, a statement in 1824; France, 1833; Sweden, 1842; Finland, 1856; Paraguay, early nineteenth century; United Kingdom, 1870-1892; Bahamas, 1878; Bulgaria, 1879; Argentina, 1884; Mexico, 1892; Australia, 1895; the Netherlands, 1901; Sri Lanka, 1907; and Belgium, 1914. The education given remained somewhat restricted in scope focusing on the provision of reading and writing skills and the acquisition of religious knowledge.

In many countries until very recently a dual system existed with separate schools for children who were able to continue towards higher levels and for those who were not. A typical example is provided by Sweden [86]. The compulsory primary school (*folkskola*) provided general training for all and lasted for six years until 1948, while the elite system of lower secondary school (*realskola*) and upper secondary school (*gymnasium*) gave academic training for students with sufficient economic and intellectual capacities. This dual system was abolished in stages. From 1984, pupils were allowed to enter the *realskola* from Grade 3 of primary school, and from 1928 they entered either from grade 4 or grade 6. Not until 1962 were the two systems abolished in favour of a comprehensive compulsory school.

Similar systems have existed in other West European countries, and in some cases remnants still exist. In the Federal Republic of Germany, it is still possible to enter a 'theoretical' secondary school stream at an early stage while other pupils remain in what is called *Hauptschule*, which does not prepare for higher education. In the United Kingdom, private schools also form an elite system, but the public schools still prepare the vast majority of candidates for higher education.

The schools created by colonial powers in their dependencies were based on their own schools. It seems that they usually served a limited purpose: to provide the trained staff of local people needed by the colonial administrations. Few attempts were made to create a general education for all, and no system able to train a sufficient staff of local teachers to act as specialists in academic fields was found. Religious schools still exist in many areas, however, providing religious and literacy education for children and adults, and in some cases also supplying advanced training for specialists. An example of the latter is the Islamic religious school system related to the Al-Azhar University in Cairo, which has trained Moslem scholars as teachers, administrators and the like for many generations.

A new way of looking at education emerged during the twentieth century; in some countries the changes took place after the Second World War, but in others it came much earlier. People began to regard education as a right to be enjoyed by everybody, and there was debate about its content.

The Universal Declaration of Human Rights mentions the right to (primary) education. This right has also been guaranteed by numerous national constitutions and laws. In Europe and in many other parts of the world, it was possible to build on the existing school systems, which were expanded and renewed. In most emerging nations, the idea of an education for all could be accepted after their independence. It is generally agreed that this right to education shall be extendedd to all people independently of sex, race, social class or language. Education is compulsory in most countries in the world. The duration of compulsory education is, furthermore, rarely stated as a number of years but in terms of age or of level of schooling achieved. Conflicting information is found.

There are countries where no compulsory education exists. The reasons vary; in some cases they are economic and financial coupled to the realization that it is not useful for a country to introduce laws which cannot be enforced, but in others, for example, Bahrain, Lebanon, Qatar and Saudi Arabia, the reason is stated to be to safeguard the freedom of people. Compulsory education exists in countries like Denmark, Ireland and Peru in so far as the person in charge of the child is responsible for the provision of a proper education; in such cases it is possible to replace education in an organized school with schooling at home. Malaysia and Tunisia state that primary education is not compulsory but a right, while the Central African Republic has made it compulsory as soon as the child enters school, though it is not a right. China and Indonesia have advanced plans for making primary education compulsory. In many countries with compulsory education, enrolments fall short of expectations.

The school systems all over the world present a very homogenous picture. The model used is mainly European in origin. Holmes [22] states that the new system in Prussia and the formulation of similar comprehensive plans in the United States and France during the eighteenth century provided the models which were later followed by other countries until recent times. The colonial powers transferred their ideas of education to their colonies and protectorates, and, after independence, these ideas were adopted by the new rulers. Not until recently have certain developing countries started to look to other traditional models as a means for the universalization of education, particularly Moslem countries like the Arab States, Pakistan and Nigeria.

The traditional European model of education had certain features which still characterize many school systems [25, 26]: (a) there are fixed ages for entering and leaving school; (b) the system provides full-time education; (c) it is teacher dominated; (d) each level is divided into grades normally corresponding to one year of study; and (e) examinations play an important part.

This model originally built the curriculum for primary education around the 'three Rs' (reading, writing and arithmetic) and religion, while the curricula for the higher stages were built around religion and classical languages. Well into the twentieth century the curricula in all schools were traditional and a definite ranking order between subjects existed in the upper school [22, p. 9]. The main aim of the lower school, besides providing certain basic skills, was to prepare students for further schooling ending in traditional university education. This had consequences for the way in which the school worked. A student who was found unable to profit from higher education in the traditional sense was prevented from entering or continuing in secondary school. Successive dropping out provided a stock of people with less extensive education who could enter the labour force. Many education systems, particularly in developing countries, still retain some of these features, as will be discussed in connection with the performance of the systems.

There are countries which have never been colonized and which do not belong within the European cultural sphere; although influenced by European culture, their educational traditions differ from the European one. An example is Japan, where there was a long educational history before the country was 'opened' to Western people around the middle of the nineteenth century. Even so, the structure of the school system is fairly similar to the European model, although there are said to be great qualitative differences between schools at the same level and keen competition to enter the 'best' ones [26, p. 84].

A new model of education has emerged during the past four or five decades [22, p. 13]. According to this model, it is necessary to question not only the organization but also the content of education. It should be an education for everybody; all subjects are of equal value if they meet a need. Most students will in the future devote themselves to some aspects of industry or commerce; a much smaller proportion will choose a traditional vocation. This means that primary school, like secondary school and the university system, must teach skills useful in practical life. A societal development with a heavy stress on social justice and a postponement of vocational choice has meant that the school system shall not force children to make early choices of their future occupation while still in school. The modern idea of primary and lower secondary school has therefore become democratic, relevant, flexible and undifferentiated.

It is felt that basic schooling shall prepare children for active life, independent of their future vocations or positions in society. The situation differs between countries, however, and in many cases regional and local needs must be taken into consideration. This has made many countries anxious to adapt their school systems to the environment, as will be discussed in connection with the goals of primary education. There are two particular points which should be mentioned already here, however, namely the new role of science and technology in basic education, and the part played by school in the preparation for productive work.

The traditional school was created during a time when little scientific knowledge existed and still less was considered useful. Natural sciences were found of interest to specialists only. This situation has changed dramatically. Modern industry, agriculture and services are based on scientific ideas and discoveries; the growth of scientific knowledge has been phenomenal; and world events and improved communications have made many scientific discoveries and their consequences — such as atomic power, new medicines, television, radio and space research — well known to everybody. Scientific literacy has become essential since it enables people to understand and cope with the world facing them. This practical value and theoretical interest means that natural sciences are considered essential parts of any curriculum. To provide the young person with an access to these sciences, it is also necessary to provide him/her with a set of tools, and technological education forms another essential element of schooling.

It is generally accepted that education shall be a preparation for active life, and all pupils must prepare for the time when they will leave school. In the traditional system, as it was organized before the Second World War, school was separated from work: the learning of a trade or another non-academic vocation usually took place outside the school after the young person had finished his formal education. In this respect, education was isolated from society. This has changed: education is viewed and usually functions as an integrated part of society. This means that training for jobs shall take place within the framework of the school. Even at the primary stage there are job-preparatory activities. On the other hand, education has become a lifelong activity which occupies part of the active life of most workers.

Besides acquiring the necessary vocational skills, children should also learn to understand productive life in the society in which they live, and they should learn to see the relationship between what they learn in school and what is needed in society. This problem has been solved in different ways in different countries, with some of the socialist ones stressing practical work outside the school as part of the timetable. In all cases, there is a felt need for a smooth transition to active life and work.

Although these tendencies in the development of primary and lower secondary education are often more pronounced in developed countries, whether socialist ones or not, they can be found in developing ones as well. In the latter case, there are great problems for the teaching of natural sciences in the form of inadequately trained teachers and a lack of other resources. In addition, the environment is different from that in a technically developed part of the world, which means that different technologies are used and taught and that different types of vocational preparation are useful. This will form a main theme in the discussion in this book.

The term 'primary education'

The term 'primary education' and its equivalents in other languages are used quite differently in different countries. In order to discuss its nature, it is necessary to get answers to the following questions:
1. At what age do children enter primary school?
2. How many grades are there?
3. Does primary education form a natural part of the education system?

In some cases, primary education is defined as a five-year cycle which children enter at the age of 5 (for example, Lebanon and India) and, at the other extreme, it can also be defined as an eight-year cycle entered at the age of 7 (Brazil, Poland, Rwanda and others). In the Soviet socialist republics, primary education covers three years from the age of 7 (the entrance age may be lower), and Mongolia also has a three-year primary cycle. It is quite obvious that aims and organization, as well as the way in which subjects are taught and innovations introduced, must differ considerably between a system where pupils are between 5 and 10 years old and one where they are between 7 and 15.

Primary education does not always form a natural cycle of the educational system nor coincide with compulsory education. In some countries the latter is of longer duration than primary school, and then the compulsory years may form the natural cycle. In other cases primary school may be divided into cycles, and some countries divide the total system between pre-primary school and university in such a way that primary school becomes a name only. It is not possible to give a complete picture of the systems in all countries, but a few examples may be useful.

In the USSR, the term primary education covers only the first three years of the general polytechnical school. Primary school may form part of either the complete (ten year) or the incomplete (eight year) secondary school, but it may also be a separate school with its own administration. In Czechoslovakia, where compulsory education lasts for ten years from the age of 6, there is an integrated eight-grade school consisting of two levels of four years each. Bulgaria has a unified eleven-year polytechnical school with transfer to technical or vocational education after grade 8. The term 'primary education' applies only to the first three years. A new system is planned with a first stage of ten grades leading to secondary and specialized education. In the German Democratic Republic, there is also a ten-year polytechnical school, called secondary school. It is divided into three cycles; the first cycle of three years; the middle cycle of three years; and the upper cycle of four years. The term 'primary education' is not used. In Hungary, compulsory education of eight years consists of a lower and an upper four-year cycle. Only the former is called primary education.

In Romania, the ten years of compulsory education are divided into: elementary education, four years; gymnasium, four years; and the first cycle of the lycée, two years. Greece has nine years of compulsory education: six years of primary school and three years of gymnasium.

In the Nordic countries there is a nine-year compulsory school divided into levels, usually of three years each. In Denmark, however, the first seven years are called *folkeskole* (primary school; literally, school for people), but there is no demarcation line between primary and lower-secondary school. In the Netherlands, a new law, effective from 1985, will integrate pre-primary and primary education to basic education for children between 4 and 12 years of age. Italy and San Marino have a basic education divided into primary (five years) and secondary (three years). In Spain, the term primary school is given two interpretations: (a) the *ciclo inicial* (two years) plus the *ciclo medio* (three years); or (b) these two cycles plus also the *ciclo superior* (three years). In the latter case, it corresponds to compulsory basic general education. In Belgium, pre-primary and primary education together form fundamental education from 5 years of age. In Malta, where secondary education is also compulsory, primary education consists of two cycles of three years each. In the United States, primary education generally lasts for between six and eight years depending on the local system. The distinction between primary and secondary education, however, has become blurred by the growth of middle schools.

Algeria has integrated her old primary and intermediate school to form basic education of nine years. Jordan has a 6+3+3 system, but compulsory education is for nine years and the second (preparatory) cycle is similar to the first (primary) one. In Egypt, there is a basic nine-year education system with two stages — primary (six years) and preparatory (three years).

Cameroon has two primary systems, the French-speaking one of six years and the English-speaking one of seven years. The Kenyan government has issued a recommendation implying eight years of primary school. In Malawi, there is an eight-year primary cycle, and students sit for examinations at the end of years 5 and 8. Nigeria has six years of primary school with an expected transition of 100% to lower secondary school.

In Chile, basic education covers eight years and is divided into two cycles of four years each. The eight years of primary education in Bolivia, on the other hand, are divided into a basic cycle of five years and an intermediate cycle of three years. In Brazil, compulsory education is for eight years. Paraguay has a primary school with six grades and distinguishes between two cycles of three years each. Basic nine-year education in Colombia is divided into five years of basic primary and four years of basic secondary education. Venezuela also has a nine-year basic school. The seven-year primary school in Argentina consists of three cycles of three, two and two grades, respectively.

THE PERFORMANCE OF PRIMARY EDUCATION

An aim of the present book is to contribute to changing some features of the primary education systems of some countries and to make these systems more responsive to the challenges of the modern world. This will be done by raising questions and pointing to weaknesses in existing systems.

The question of the functions and functioning of primary education systems will be discussed in this and the following chapters. The concept of the performance of an education system will be discussed here, and a model of factors influencing such performance will be presented. The major part of the text will deal with indices of performance such as scores in standardized tests, enrolment ratios, dropping out of school, repetition of grades and discrimination in the system. Finally, a discussion of problems facing primary schools, particularly in developing countries, will summarize the discussion.

Indicators of performance in an education system

The question of what is meant by 'quality of education' has been discussed by many authors without any real answer being presented. This book will not try to enter into this discussion. Instead, it has been found more rewarding to try to contribute to the discussion of another essential question: what is meant by the performance of an education system?

Here it is possible to give at least a provisional answer. By the performance of an education system is meant the extent to which it reaches the goals and meets the objectives of this system.

Such goals and objectives can be of several kinds [91, p. 116]:
- *primary* goals and objectives refer to the behaviour of the system as such or of the individuals which are in or have left the system;
- *secondary* goals and objectives are those of products of the education given in the system: knowledge, skills, attitudes, values, etc., which are assumed to enable the system to reach the primary goals and objectives; and
- *tertiary* goals and objectives are those of the immediate characteristics of the system itself: curriculum content, administration, organization, teaching methods, equipment and so on; these characteristics aim to allow the system to reach secondary goals.

To be able to tell whether a system reaches its goals and objectives, it is necessary to consult adequate criteria, which will accordingly be primary, secondary or tertiary.

Primary criteria indicate whether the behaviour of the system or that of its individuals is in accordance with objectives: flows of students into the system; equity within the system; flows of students through the system (dropping out, repetition, graduation, absenteeism, etc.); distribution by type of study; vocational choices of students; vocational behaviour of previous students; social behaviour of previous students; and so on.

Secondary criteria measure whether the products of the system are of satisfactory quality: knowledge of students and graduates; skills of students and graduates; attitudes of students and graduates; and so on.

Tertiary criteria measure whether the characteristics of the system itself are in accordance with goals and objectives. They answer questions like: are the organizational features correctly interpreted?; and so on.

That the construction of some of the criteria may cause very considerable difficulties is only too obvious, particularly since few countries present objectives in such a way that they can easily be translated into criteria. Some criterion measures are not readily available either, for example those which concern the behaviour of students after they have left the system. Nonetheless, a great deal of educational research has been devoted to studies of performance according to some of the types of criteria mentioned above.

In any analysis of the performance of an education system, a selection must be made of criteria. The problem becomes simpler when an author wants to study whether certain phenomena exist compared with a situation where he wants his results in quantitative terms. Most authors who have discussed quality of education have limited themselves to treating certain effects of schooling on student behaviour or on products like knowledge, skills and attitudes. These are, undoubtedly, important aspects. Also when the performance of the education system is studied, it is of importance to consider what happens to individuals: does the system attain the stated goals with respect to their training? There are other goals and objectives which should also be considered; for example, those which have to do with societal development, since they are equally important.

Unfortunately, it is not possible to present data on more than a few aspects of educational goals in this book. Some attempts have been made to measure student performance in some subjects in a few countries and to make an international comparison, thus enabling a discussion of the relative performance levels of a few school systems.

More has been done in the area of students flows, which allows statements about the extent to which the systems reach goals of efficiency, universalization of primary education and equity. As examples of such studies those by Elgvist-Saltzman [12], and Lindgren-Hooker [42] can be mentioned. The former studied the flow of students through a university system, while the latter investigated flow models applied to statistical data from a number of developing countries. A long series of studies have concerned the relationship between line of study and vocational choice. Analyses of selection and streaming of students can be found in handbooks of educational research.

To state that an education system shows inadequate performance in some respect is often of little value unless something can be said about the causes. These can be found in society at large, in the school and its environment, in the families which send children to school, in the characteristics of the students and the non-enrolled children who should attend school, and in the curricula and other elements of the school system which may influence its quality.

It can be shown that these causes can be organized into a so-called 'path model', which will therefore be used in the following discussion. A path model is characterized by the fact that it divides all influencing factors into groups $1,\ldots x,\ldots n$, so that the factors in group x can be imagined to influence those in groups $x+1, x+2, \ldots, n$, but cannot be influenced by the latter. All influences between groups or sets of factors are therefore in only one direction.

A picture of such a model is given in Figure 2. In the model, circle 1 represents societal factors; circle 2, factors in the schools and their environments; circle 3, factors related to home and parents (including their socio-economic status and education); circle 4, the characteristics of the pupils and non-enrolled children which cannot be considered an output of the education system; circle 5, the characteristics of the education system which are studied, for example curricula; and circle 6, criteria measuring the performance of the system.

An arrow in the model represents influences of one set of factors on another set. To take a single example, the factors in circle 5 as a group are influenced by factors in circles 1 to 4. The characteristics of the education system should be adapted to factors like the geography of the country, the linguistic and cultural characteristics of the population, the socio-economic situation and educational needs of society, and so on. They should

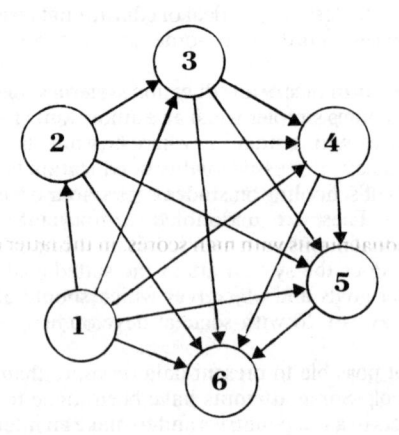

FIGURE 2. Path model of factors influencing educational performance

Contexts of science and technology education

also be adapted to the quality of the schools, to the backgrounds of the teachers, to the standing of the schools in the community, etc. They should take into account factors like the literacy level and other educational status of parents as well as their cultural, religious and socio-economic situation. Finally, they should consider the fundamental characteristics of the children, for example, their intelligence, attitudes acquired outside school, language and so on.

A great deal of research has gone into the study of the effects of individual factors on the performance of an education system. Much of this is relevant to the present discussion, and examples will be given in the text below. It would also be possible to study the whole complex of factors influencing an aspect of the quality of an education system. As far as the authors know, no such attempts at a global approach have been made.

Comparisons of student performance in different countries

A great deal of the educational debate going on in the world concerns the 'standard' of the schools, that is, the extent to which students are able to reach a 'performance' (knowledge, skills, etc.) level which is considered adequate or desirable. This debate is often very emotional, and people without special knowledge of educational measurement take part. Around 1960, the launching of the first man-made satellite led to a discussion of the standard of schools in the United States, which was assumed to be inferior to that in the USSR and other countries. Recently, this discussion seems to have started up again and has led to the creation of governmental committees with the task of suggesting steps for improvement. During the long period of educational reform in Sweden which started around 1955, critics maintained that the new school would lead to a lowering of educational standards. A number of large studies were carried out to study whether this was really the case [79, 10]. In countries where a strict examination system exists and students are forced to repeat grades, educators maintain that this is a means to safeguard the standard of the schools.

An international organization of research institutions, called the International Association for the Evaluation of Educational Achievement (IEA), has carried out a series of comparative studies in different countries, first in respect of mathematics [27, 58], then in six subjects including science [59], then again in mathematics, and once more in science [45].

Researchers engaged in the IEA projects are anxious to point out that it is extremely difficult to make comparisons between different countries: curricula differ, values differ, the environments in which the children grow up differ, and so forth. The authors tried to take the latter into consideration when constructing the tests.

A few results referring to performance in science among 14-year-old pupils may be quoted. The first data column in Table 1 gives the percentages of all these pupils who scored more than 45 points in the first science test. The second data column indicates the percentages of all schools in different countries whose mean scores are below the mean for the lowest scoring school in Japan. In the former case, a high percentage thus indicates many individual pupils with high scores, in the latter case, it means that many schools show low scores.

Independently of the way in which the data have been treated, it is obvious that pupils in certain developed countries, notably Japan, Hungary, Australia and New Zealand, perform very well. On the other hand, those in a number of developing countries — Thailand, India, Chile, the Islamic Republic of Iran — show low scores on this test.

The researchers involved tried to find explanations for the differences between the countries, and their analyses revealed the following main causes [26, p. 108 and following].

1. Amount of time devoted to the subject: number of years learning science; number of hours per week in the timetable; time spent on homework, etc.;
2. The curriculum: content, delivery system;
3. Teacher qualifications; and
4. Classroom climate, as measured by means of an attitude questionnaire.

These factors belong in circle 5 in Figure 2. All are such that they can be influenced by educational authorities. Obviously, there are other factors, not studied by the IEA projects, which might influence student performance in science.

TABLE 1: Performance level in science tests of 14-year-old pupils in different countries

Country	Percentage of pupils over 45 points	Percentage of schools below standard
Japan	34.2	0.0
Hungary	26.2	2.1
Australia	26.2	7.2
New Zealand	15.9	8.3
Sweden	10.2	9.3
Federal Republic of Germany	10.7	12.6
United States	11.7	20.3
Scotland	13.3	37.0
Finland	6.9	36.4
England	13.2	45.7
Belgium (Flemish-speaking part)	3.4	41.4
Netherlands	3.3	44.6
Italy	3.2	63.0
Belgium (French-speaking part)	1.3	61.9
Thailand	0.9	62.8
India	1.4	87.9
Chile	1.5	90.9
Iran	0.0	100.0

Source: [45] & [26]

Enrolment in school

An important indicator of the performance of an education system is enrolment in primary school. This will be examined below; the enrolments in secondary school are also shown for comparison. No school system can claim to satisfy the demand unless children enter school and stay there as pupils for a substantial period of time.

Statistical data on educational development in the world were published in connection with the thirty-ninth session of the International Conference on Education [35, 36]. They show a rapid increase in gross enrolment figures during the past twenty-two years. The total enrolment in primary school in the world in 1960 was estimated to be 340 million. In 1982 it was 575 million. In 1960 the enrolment was distributed so that there were 124 million primary school pupils in schools in developed countries and 216 million in developing ones. In 1982, the corresponding figures were 125 million and 449 million, respectively.

The secondary school enrolment in the world was 83 million in 1960, 237 million in 1982. In 1960, there were 46 million secondary school students in developed countries, 37 million in developing ones; in 1982 the corresponding figures were 81 million and 156 million, respectively. Also here, the largest part of the increase refers to developing countries. Unesco has estimated rates of increase, as shown in Table 2.

TABLE 2: Rate of increase in enrolment figures, 1960-1982, per cent per year

	1960-70	1970-75	1975-82
Primary school			
Developed countries	1.0	−1.0	−0.6
Developing countries	3.7	5.1	1.7
Secondary school			
Developed countries	4.3	2.4	0.2
Developing countries	7.9	8.4	4.1

The rapid expansion of primary school enrolment is explained almost entirely by the efforts made in developing countries. The developed ones had all reached practically universal primary education by 1960, and the stagnating birth figures since the Second World War stabilized enrolment figures. In developing countries, where there was an unmet need for education, the enrolment has continued to increase during the whole period. The increase reached a peak during the 1970-75 period. Since then enrolment has slowed down. The very urgent need for improved educational facilities is no longer felt in all these countries, and the economy in many areas limits their abilities to invest in education. The development of enrolment is now probably slower than the growth of the population in most developing countries.

At the secondary level, enrolment figures in developed countries have continued to expand, even if it seems as if a saturation point will soon be reached. In developing countries there has been and still is a rapid increase in secondary school enrolment.

These figures provide a very inaccurate picture of the situation: they do not take the population growth into account; primary and secondary education are of different durations in different countries; and repetition and dropping out tend to prolong or shorten the period of stay in school, which influences the enrolment figures.

The best way of measuring the effects of educational effort on enrolment would be to present age-specific enrolment rates. These can be found by dividing the number of enrollees who belong to a particular age cohort by the size of the corresponding population age group [88, 42]. Unfortunately, few countries provide data in a form that would permit the present authors to compute these rates.

Instead, so-called gross enrolment rates have to be used. This type of rate is computed by dividing the total enrolment in a cycle, regardless of the age of the students, by the population age group which, according to national regulations, should be enrolled at that level. The organizational structure of the school system of the country must thus be considered. The rate is defined for each country and each level separately, and it refers to quite different phenomena depending on the ways in which education is organized. It is not possible to compute gross enrolment rates for regions but, with due precaution, averages can be estimated.

The gross rates suffer from considerable weaknesses. Population data are often unreliable, quite often underestimated. It is common to find that children enter school at a later date than the one stipulated in regulations. Repetition of grades, which is an extremely common phenomenon, means that a large number of students are above the official age limit. In spite of this, these rates provide an indicator of the enrolment situation in various countries.

Table 3, taken from a statistical summary prepared by Unesco [36], shows the gross enrolment rates in primary school for 104 developing countries.

TABLE 3: Gross enrolment rates in primary school in developing countries, 1982

Region	Gross enrolment rates								
	Below 50%	50-59%	60-69%	70-79%	80-89%	90-94%	95-99%	100%	
Africa (45 countries)	Burundi Chad Ethiopia Gambia Guinea Mali Mauritania Niger Senegal Sierra Leone Somalia Upper Volta	Sudan	Benin Liberia Malawi Uganda	Cen. Afr. Rep. Egypt Ghana Rwanda Zimbabwe	Equatorial Guinea Côte d'Ivoire Morocco	Nigeria Zaire	Algeria Kenya Zambia	Botswana, Cameroon, Cape Verde, Congo, Gabon, Guinea-Bissau, Lesotho, Libyan A.J., Madagascar, Mauritius, Mozambique, Swaziland, U.R. Tanzania, Togo	
(100%)	17.7%	4.0%	6.8%	15.0%	5.9%	25.0%	11.2%	14.4%	
Latin America & the Carribbean (26 countries)			Haiti	Guatemala	Bolivia El Salvador Honduras	Trinidad & Tobago Nicaragua	Brazil Guyana Jamaica	Argentina, Barbados, Chile, Colombia, Costa Rica, Cuba, Dominican Rep., Ecuador, Mexico, Panama, Paraguay, Peru, Puerto Rico, Suriname, Uruguay, Venezuela	
(100%)			1.6%	2.0%	4.9%	1.0%	39.5%	51.0%	

	Afghanistan Bhutan Yemen A.R.	Bangladesh Oman Pakistan	Saudi Arabia	Burma India Yemen Dem. Republic	Malaysia Nepal	Kuwait Lao People's Dem. Rep. Sri Lanka Thailand	Bahrain, China, Indonesia, Iraq, Islamic Rep. Iran, Rep. of Korea, Lebanon, Mongolia, Philippines, Qatar, Singapore, Syrian A.R., Turkey, U.A. Emirates, Viet Nam
Asia (31 countries) (100%)	1.4%	7.8%	0.4%	28.3%	1.2%	3.1%	57.8%
Oceania (2 countries) (100%)		Papua New Guinea 80.5%					Fiji 19.5%
Total (104 countries) (100%)	15 countries 3.8%	1 country 0.6%	9 countries 6.9%	7 countries 2.9%	9 countries 21.7%	6 countries 4.9%	10 countries 9.1%	47 countries 50.1%

Source: [36]

It can be observed that about half of the countries show rates well below 1, and quite a few give rates below 0.50. It is also obvious that many of the countries suffer from an educational situation which is even less satisfactory than these rates indicate. Countries like Chile, Colombia, Gabon, Lebanon, Mauritius, Paraguay, Peru, the Syrian Arab Republic, the United Republic of Tanzania and Vietnam, which show enrolment rates above one, report considerable difficulties providing primary education for all children.

The development of these rates since 1960 has also been presented by Unesco. Table 4 gives the data for major regions. It is obvious that some of the rates given are greatly over-estimated, for the reasons mentioned. Those for Northern America (essentially the United States and Canada) indicate the size of the error. Nonetheless, the development has been rapid.

Although the enrolment rates for primary school thus often seem encouraging and although the world situation has improved very considerably since 1960, the situation is far from satisfactory. A crude estimate, taking into account the fact that gross enrolment rates overestimate the proportions of children who attend school, seems to indicate that as many as between one-fifth and one-third of all primary school-age children in developing countries (China not included) do not attend school at all. This shows that there is still a great deal to be done before primary education becomes generalized in all countries. The last decade has also witnessed a decline in the educational effort, as shown by stagnation in the rates.

Enrolment figures in school are a product of several factors, of which the most important one is the number of children entering school. Many countries report that a large number of children never enrol in any school.

Reasons for non-enrolment of school-age children have been mentioned by a large number of countries in their reports to the International Conference on Education. Such reasons can be found in any of the groups of factors mentioned in the model above.

The *socio-economic situation of the country* may mean that it cannot afford to extend schooling to all children. It is referred to by countries like the Central African Republic, Colombia, Egypt, Ethiopia, Mauritius, Morocco, Pakistan, Senegal, Zambia and Viet Nam. Chile reports that limited resources make it difficult to apply, on a large scale, the educational measures that benefit the sectors of the population that need education most, and Jamaica and Paraguay report a lack of funds for handicapped children.

Geographical problems mentioned include a widely dispersed population, transportation problems and bad communications. These problems become particularly pronounced with countries consisting of a large number of islands, for example, Maldives, Seychelles and Tonga. Guyana mentions its forested and mountainous terrain. Internal migration is a common phenomenon mentioned by, for example, the Syrian Arab Republic and Poland. The geographical, ethnic and linguistic heterogeneity of the country is mentioned by Mexico, and their rapidly growing population figures by Brazil, Egypt and Uganda.

Among *problems in the schools,* a large number of countries mention a lack of facilities. Tonga, Uganda and Zambia observe a lack of housing for staff. Two-shift schools are specifically mentioned by Egypt and Jordan, but are found in many other countries. That qualified teachers cannot be found, particularly in rural areas, is a very common phenomenon in developing countries. A dearth of good curricula, suitable textbooks and other materials is a reason reported by Indonesia, Jamaica, Madagascar, Nicaragua, Nigeria, Pakistan, Peru, Rwanda, the United Republic of Tanzania and Zambia. A particular problem is met in Zimbabwe, namely the integration of white and black pupils in school, which meets with resistance.

The socio-economic *situation of the family* is mentioned as a reason for the child's not

TABLE 4: Adjusted enrolment rate (per cent), 1960-1982

Area	1960	1965	1970	1975	1980	1982
Primary education						
Europe and USSR	103.0	103.8	103.8	101.0	103.7	105.0
North America	116.9	119.1	117.4	123.4	122.4	119.8
Africa	44.1	51.8	56.6	66.7	78.1	80.7
Latin America and the Caribbean	73.1	82.2	92.4	98.0	103.9	104.7
Asia (except China and D.P.R. Korea)	79.4	85.1	80.3	94.2	91.5	90.3
Oceania	101.5	102.2	103.4	98.0	100.5	100.9
Arab States	49.3	60.2	63.6	73.9	81.0	83.2
Developed countries	105.9	106.4	106.3	105.3	106.6	107.0
Developing countries	72.2	79.7	77.8	90.3	90.8	91.5
World	81.9	86.6	84.8	93.6	93.9	94.3
Secondary education						
Europe and USSR	45.6	57.0	63.5	71.9	76.4	78.7
North America	70.4	78.0	82.5	83.4	83.1	86.0
Africa	5.1	7.8	11.0	15.5	21.0	24.6
Latin America and the Caribbean	14.3	19.4	24.9	35.4	44.2	47.5
Asia (except China and D.P.R. Korea)	23.1	25.3	30.2	38.3	43.3	41.3
Oceania	52.8	61.4	68.1	72.0	70.6	71.5
Arab States	10.1	15.9	21.1	28.5	36.8	41.2
Developed countries	54.8	65.2	69.9	75.6	78.3	80.5
Developing countries	16.7	19.2	25.1	33.1	39.1	40.4
World	27.3	31.6	35.0	42.8	46.9	48.0

Note: Arab States have also been included under both Africa and Asia

being sent to school by countries like Bahamas, Egypt, India, Jamaica, Mexico, Nicaragua, Pakistan and Thailand. Disorganized and broken families may also affect enrolment, according to Peru. Social and health problems are negative factors, according to India, Jordan, Pakistan and the Syrian Arab Republic. In some areas it is found that parents are reluctant to send children to school, particularly in the case of girls, for example, in Gabon, Nigeria, the Syrian Arab Republic, Thailand, Turkey, the United Republic of Tanzania and Uganda. Rwanda states that education in the mother tongue is looked upon negatively, while Angola finds that teaching in a foreign language poses difficulties. A problem encountered in a large number of countries is child labour; children are needed to work in the home or on the family farm, to work away from the home, or to look after younger brothers and sisters. Bangladesh finds that parents are sometimes reluctant to send children to school because it alienates them from the land.

It is estimated that if the development of primary-school enrolment observed during the period 1960-82 continues at the same rate till the end of the century, there will be more than 1,000 million pupils in the world, and the goal of universalized primary education would be reached in a few more countries [31]. Plans, laws and other documents also indicate a willingness on the part of most countries to continue their efforts in the field of primary education. Education is still considered both a human right and a tool for the economic development of the country and for the creation of a social environment in which different groups can live and work together.

On the other hand, the further expansion of primary education has met with difficulties. Few developing countries seem willing to devote a larger and larger share of their GNP to education. It is not likely that all countries can reach the goal of universal primary education in the foreseeable future, and still fewer would be able to use this level of education as an efficient tool for socio-economic development unless essential innovation is undertaken in the system. Such innovation may call for changes in organization, curriculum content, teaching methods, teacher training and other conditions. To be efficient, it must be such that education becomes relevant to the environment in which it is given and is perceived as important to the individual's future.

Primary education is the main means for combating illiteracy, adult education playing only an auxiliary part. In 1970, there were 760 million illiterates in the world, which constituted about 33 per cent of the adult (15 years of age and older) population. Of the illiterates, 731 million were found in developing countries, that is, 48 per cent of the combined adult populations of these countries. The illiteracy rate was 71 per cent in Africa and 73 per cent in the Arab States, lower elsewhere. The situation has evolved slowly. It is estimated that there were 857 million illiterates in 1985, but the higher figure is due to a population increase. Of the illiterates, 838 million are found in developing countries, where they represent 37 per cent of the adult population. Again, the highest rates are found for Africa (54 per cent) and the Arab States (57 per cent).

These data should be treated with care. They are rarely based on careful measurement of skills but usually on subjective estimates made during a census. They do not take into consideration the fact that the world is getting more and more complicated, calling for better and better literacy skills. Although there seems to be an improvement, the ranks of illiterates continue to grow as many children do not enrol in and pass through primary school. With the levelling off of the catching power of the school system, the solution to the illiteracy problem is still not in sight.

Discrimination in education

Discrimination in education means that a group or an individual is given less opportunities to enrol in a school, less possibilities of following education in a certain specialization or along a certain stream, or less opportunities to profit from schooling, than other groups or individuals. It means that the group or individual loses opportunities, but it also means that society is left without the benefit of the education that could have been given.

Judging from official documents and laws, discrimination should be a rare phenomenon. In a questionnaire on the application of the convention against discrimination in education, answered during 1983, Unesco inquired about practices in national education systems which might constitute discrimination. With a couple of exceptions, the Member States were very adamant on this point: there are legal provisions which rule out all kinds of discrimination. One exception was Namibia, which reported discrimination on the basis of race. Seven times as much was spent on a white child's education as on a black or 'coloured' child's education; only 83 per cent of black children were enrolled in primary and only 16 per cent in secondary education.

The other exception was a report from Nigeria about a quota system adopted for admission to federal secondary schools. This system may prevent certain able students, particularly from southern states, from entering, while other less able ones are admitted. Similar 'positive discrimination' systems are found in other countries as well, even if they were not reported as discrimination. Sweden and many other developed countries spend comparatively more on the education of disadvantaged groups than on others, and in some cases positive steps have been taken to induce males and females to enter lines of study where their sex is under-represented. India reports that incentives in the form of meals, school uniforms, scholarships, etc., are given to girls. In many countries, special attention is paid to isolated and backward areas so as to improve their educational situation. These measures to create an equitable balance between the opportunities of different groups are not considered as discrimination.

No country reports any discrimination on the basis of sex. Saudi Arabia has separate education systems for males and females but states that these systems offer equivalent access and equivalent education. Jordan, Kuwait, Sudan and the Syrian Arab Republic report separate schools for boys and girls, but always with the same opportunities and equivalent quality and under the control of the same authority. A long series of countries mention that one-sex schools do exist together with co-educational ones: Australia, Belgium, Bermuda, Denmark, Ghana, India, Indonesia, Ireland, Japan, Sierra Leone, Spain, Sweden, Trinidad and Tobago, and so on. No discrimination is said to be caused by this since the schools are of the same quality and provide equal facilities.

A look at statistical data on educational enrolment by sex tells quite different a tale, however; there are great differences between the sexes. Table 5, abstracted from the Unesco statistical summary [36], shows the proportions of female students in primary and secondary school in 1960-82. To interpret these data, it should be noted that the figure 33 per cent means that there were twice as many boys as girls in school; 40 per cent, that there were one-and-a-half times as many.

The situation has undoubtedly improved. In most parts of the world, including Latin America and the Caribbean, there are approximately equal numbers of boys and girls in primary and secondary schools. Asia and Africa provide exceptions. In these continents, there are about 25 per cent more boys than girls in primary schools and 50 per cent more boys than girls in secondary schools.

There is another type of educational sex discrimination, which does not appear in this table (nor in any other tables) and which is much less often discussed, namely the one which appears in the choice of *lines* of study. All over the world, females are found almost entirely in streams which have a humanistic, educational or social inclination, while there are very few studying scientific and technical subjects. This is related to and has an effect on their future employment situation. In a country like Saudi Arabia, the vocational choices of women are regulated by means of laws, which means that their

TABLE 5: Female enrolment (percentage of total enrolment), 1960-1982

Area	1960	1965	1970	1975	1980	1982
Primary education						
Europe and USSR	49	49	49	49	49	50
North America	49	49	49	49	49	49
Africa	36	39	40	42	44	44
Latin America and the Caribbean	48	49	49	49	49	49
Asia (except China and D.P.R. Korea)	39	40	41	43	43	43
Oceania	48	48	48	48	48	48
Arab States	34	35	36	38	41	42
Developed countries	49	49	49	49	49	49
Developing countries	39	40	41	44	44	44
World	43	43	44	45	45	45
Secondary education						
Europe and USSR	48	49	49	51	51	52
North America	50	50	50	50	49	49
Africa	29	30	32	35	38	40
Latin America and the Caribbean	47	48	48	48	50	50
Asia (except China and D.P.R. Korea)	35	35	36	38	39	39
Oceania	48	48	47	48	49	49
Arab States	26	27	30	34	37	38
Developed countries	49	49	49	50	50	50
Developing countries	32	33	35	38	39	40
World	41	42	42	43	43	44

Note: Arab States have also been included under both Africa and Asia

studies are restricted. In many other countries, certain vocations, like religious offices, can only be held by men. But even in Scandanavian countries, where all vocations are open to women and considerable efforts have gone into making people choose vocations dominated by the other sex, the choices of streams are still conditioned by traditional practices.

This problem is related to attitudes in society, often to superstitions or to situations which no longer exist, but which resist change. Unesco has attacked it, and its Medium-Term Plan states:

There are numerous historical, social, economic and sometimes cultural reasons for this situation, such as low school attendance during the colonial period, the inadequacy of family resources, numerous and restricting domestic obligations, inadequate facilities and the work thereby entailed, early marriages and early pregnancy, and a certain wariness of the attitudes induced by attendance at modern schools. This also leads to a high drop-out rate by girls and to limited participation by women in adult literacy programmes. One other remark that must be made is that, where illiteracy has virtually disappeared, certain socio-cultural behaviour patterns — whether family attitudes or biases induced by the education system — restrict the access of girls and women to certain types of training, particularly in the scientific, technical and management fields. The resulting prejudices and inequalities continue in many countries to be the source of numerous discriminatory practices against women regarding access to certain professions or real career prospects.

Many countries report difficulties providing education for children in remote geographical regions: Algeria, Australia, Colombia, Denmark, Egypt, Ethiopia, Malawi, Maldives, Morocco, Mexico, Nepal, Pakistan, Seychelles, the United Republic of Tanzania and others. The difficulty may be caused by long distances and a scattered population, for example in arid areas, by mountainous terrain, as in the Andean countries and Algeria, by the country consisting of islands, or by the climate. Colombia states that economic, nutritional, communication and family factors, and a lack of school facilities, tend to cause inequalities in educational opportunities; and the Syrian Arab Republic states:

Despite the facilities and services which the State has made available to the people, however, a number of difficulties obstructed the execution of the commitment to compulsory primary education. Some of these difficulties are economic, such as low income of some households. Others are social, such as the customs and conventions regarding the education of women, and still others are cultural in nature. Some difficulties are related to health, such as physical, psychological and learning handicaps, while others are related to the prevalence of nomadism in some areas of the republic.

Rural areas are always at a disadvantage. Up till now, schooling has to a large extent been urban centered, particularly in poor countries, and most of the children of primary school age who do not attend school (about 150 million) are from rural areas. In Egypt, to take a single example, primary education reaches almost all children in cities but less than 75 per cent in rural areas. The disadvantage of being born in rural surroundings does not depend only on the provision of facilities and teachers, and it has little to do with the fact that people do a particular type of work. To a large extent, the poverty of the community and the families, the need felt for the children's work, and customs and traditions prevalent in the rural society are responsible. This has been mentioned by several countries.

Other areas which are often reported as lacking proper educational facilities are the slum districts (shantytowns, *bidonvilles*) outside the main cities. This population is mostly very poor, unstable and ignorant; population registration is often inadequate; and the areas are very densely populated.

Several countries report great efforts to create educational facilities in rural areas, for example in the form of one-teacher schools, incomplete schools, mobile schools, correspondence education and so on. In spite of these efforts, the situation often remains inadequate. The school system itself is usually geared to urban life, and there are still many traditional, cultural and socio-economic factors which cause discrimination.

There are other specific groups of people who live in disadvantageous circumstances and who often find themselves suffering from educational discrimination: handicapped people, migrants, refugees, linguistic minorities, and groups with particular cultural characteristics. They live among a majority of people with other 'normal' characteristics, and for this reason they require special provisions to meet their educational needs. The education of these groups is often extremely demanding. Handicapped pupils require prolonged education and special teaching; migrants may require mobile schools and they, as well as refugees and linguistic minorities, may need teaching in their mother tongue and remedial teaching to catch up with other pupils; and linguistic and cultural minorities may demand their own schools or classes or specially trained teachers. All this is costly, and it may be hard to obtain the teachers, books and other necessary materials. Some developed countries have made great efforts to solve these problems and may even have succeeded in integrating the specific groups into their societies. Poor countries are rarely able to afford to take all the necessary steps.

To a large extent the discrimination found in education is due to factors beyond the control of the education system itself. The economic situation of the country may prevent it from building schools in very sparsely populated areas and from providing special education to all groups with particular needs. Family poverty and the economic situation in the local community may prevent parents from equipping their children for school and may make them need their presence as workers. About this, little can be done, of course. In other cases, action is possible but whatever efforts are made must be seen in a long perspective. Attitudes towards the role of women, traditions related to the behaviour of girls, and values attached to education and particular types of work influence the way the school is functioning and may cause discrimination. They can be influenced by educational means, but this takes a long time, perhaps generations.

There are, however, factors within the school system itself which may cause discrimination. Nearly everywhere, the school systems are modeled on the ones found in the Western world. They are geared to modern life. They are built on the idea of making the child able to participate in a modern society. Furthermore, one of their main aims is the preparation for further (theoretical) education. Very little time is devoted to traditional life styles, and few attempts have until recently been made to adapt curricula and organizations to local conditions. This has meant that the groups which have experienced discrimination, for example women, rural inhabitants and 'marginal' groups, have felt that they get little profit from attending school.

Wastage in education

Everywhere it is assumed that a child should spend one year in each grade in school and continue through the school cycle until graduation. Deviations from this ideal pattern are extremely common however: many children spend more than one year in each grade, and they may drop out of school prematurely, either never to return or to return after some time. In all cases, the State or the school authorities have had to spend resources and devote efforts to the education of these children without feeling that they have obtained the full benefits from this investment. Deviations from the 'normal' patterns of one grade per year and continuation till graduation are assumed to mean that resources are wasted. Unesco therefore uses the term 'wastage' as a term for such deviation.

There are several indicators of such wastage in education. The most important ones, the only ones which will be discussed here, are repetition and dropping out. It should be mentioned, however, that there are others, for example, absenteeism for prolonged periods without the child leaving school permanently, and delayed entry into school.

Repetition is the phenomenon when a student is found in the same grade during two consecutive years. A common measure of the frequency of repetition is repetition rate. It is computed by means of the following formula:

$$r_x^a = \frac{R_{x+1}^a}{E_x^a},$$

where r_x^a is the repetition rate for those who were in grade a in the year x; R_{x+1}^a the number of repeaters of grade a in the year $x + 1$; and E_x^a the total enrolment in grade a in the year x.

Dropping out is the phenomenon when a student leaves school entirely without returning. Several measures are used to indicate the frequency of this phenomenon. A measure of the tendency of a school system to lose students is the dropout rate. It is defined by means of the following formula:

$$d_x^a = \frac{D_x^a}{E_x^a},$$

where d_x^a is the proportion of students in grade a in the year x who dropped out during or at the end of this year; and D_x^a is the number of such dropouts, D_x^a can be estimated as:

$$D_x^a = E_x^a - P_{x+1}^{a+1} - R_{x+1}^a,$$

where P_{x+1}^{a+1} is the number of the students who were promoted from grade a to grade $a + 1$ at the end of grade a.

It is also possible to measure dropping out by computing so-called survival rates, which is done in the following way. An estimate is made of what proportion of the students who were in grade 1 in the year x will ever reach grade 2, grade 3, grade 4 and so on. Some of these students will reach grade 2 after one year — those promoted without repetition; others will reach it after two years — those repeating once before being promoted; and so on. In its publications *A summary statistical review...* [36] and *Evolution of wastage...* [33], Unesco has presented data on wastage in the world.

Repetition data are given for each country and are in the form of repetition rates, thus easily interpretable. They are reproduced in Table 6. The data refer to the year 1980 or 1981 and are thus a few years old.

The rates were low in most developed countries but also in countries like the Republic of Korea, Malaysia, Seychelles, the Sudan and Zimbabwe. Very high rates, around or even more than 30 per cent, were found for certain African countries.

One of the items in the questionnaire sent to Member States of Unesco in connection with the thirty-ninth session of the International Conference of Education dealt with repetition. Among the average percentage rates reported, the following ones can be mentioned: Chile, often over 10% (even higher in rural schools); Ethiopia, 13%; Jamaica, 9% in grade 6; Jordan, 4%; Paraguay, 16—18%; Rwanda, 15%; Senegal, 12—16% (36% in the highest grade); Sri Lanka, 12—13%; Thailand, 11%; and Uganda, 10-15%.

TABLE 6: Percentage of repeaters in primary education, boys and girls

Region	Below 5%		5-9.9%		10-14.9%	
Africa	Sudan	0.0	Kenya	6.9	Algeria	10.2
	Zimbabwe	0.0	Egypt	7.9	Liberia	11.5
	Seychelles	0.8	Libyan Arab		Swaziland	11.8
	United Republic		Jamahiriya	9.2	Gambia	12.9
	of Tanzania	1.2	Uganda	9.6	Niger	13.8
	Zambia	1.9			Mauritania	14.0
	Ghana	2.0			Rwanda	14.4
	Botswana	4.6				
Latin America	Guyana	3.6	Cuba	6.8	Grenada	10.8
and the	Jamaica	3.9	Costa Rica	7.4	Ecuador	11.0
Caribbean	Trinidad and		Argentina	7.5	Panama	12.4
	Tobago	3.9	El Salvador	8.8	Chile	12.6
			Mexico	9.9	Paraguay	14.1
			Venezuela	9.9		
Asia and	Japan	0.0	Kuwait	6.2	Iraq	10.2
Oceania	Rep. of Korea	0.0	Viet Nam	6.9	Thailand	10.2
	Malaysia	0.0	Syrian Arab		Sri Lanka	10.8
	New Zealand	0.0	Republic	8.2	Oman	12.3
	Cyprus	0.6	United Arab		Bhutan	12.9
	Norfolk Island	1.3	Emirates	8.5	Brunei	13.2
	Mongolia	1.9	Indonesia	8.8	Bahrain	13.7
	Philippines	2.4	Tonga	9.2	Saudi Arabia	13.9
	Hong Kong	3.6	Qatar	9.5	Afghanistan	14.8
	Jordan	4.0				
	Singapore	4.0				
	Fiji	4.1				
	Kiribati	4.1				
	Solomon Islands	4.9				
Europe and	Denmark	0.0	Luxembourg	6.1		
the USSR	Norway	0.0	Spain	6.9		
	Sweden	0.0	France	9.2		
	United Kingdom	0.0				
	USSR	0.3				
	Czechoslovakia	0.9				
	San Marino	1.1				
	Greece	1.3				
	Italy	1.3				
	Yugoslavia	1.6				
	Bulgaria	1.7				
	Federal Rep.					
	of Germany	1.9				
	Malta	2.0				
	Switzerland	2.2				
	Hungary	2.4				
	Netherlands	2.4				
	Austria	3.1				
	Poland	3.2				

15-19.9%		20-24.9%		25-29.9%		30% and above	
Sierra Leone	15.0	Zaire	20.1	Mali	26.7	Central African	
Senegal	15.7	Guinea	21.9	Congo	27.9	Republic	34.8
Lesotho	16.4			Cape Verde	28.0	Gabon	34.8
Burkina Faso	16.4			Guinea-Bissau	28.1	Angola	36.0
Malawi	17.4			Mozambique	28.7	Togo	36.8
Benin	18.2			Burundi	28.8	Chad	37.6
Côte d'Ivoire	19.0			Morocco	29.0	Sao Tome and	
Tunisia	19.6			United Rep. of Cameroon	29.9	Principe	46.6
Uruguay	14.9	Brazil	20.4	Suriname	25.8		
Nicaragua	15.3						
St. Pierre and Miquelon	15.8						
Honduras	16.2						
Guatemala	16.7						
Haiti	17.8						
Dominican Rep.	18.0						
Peru	18.5						
New Caledonia	15.6						
Bangladesh	17.8						

Portugal	16.6
Belgium	19.0

Source: [33]

Even highly developed countries without dropping out may report considerable repetition, for example, Austria, Belgium (about 8%), France (7—13%) and Poland.

Dropping out is also reported by Unesco [33, 36], but in the form of survival rates. Table 7 gives a summary of these data for eighty-eight countries combined into groups or regions. The rates given are derived by the construction of cohorts for each region, using regional promotion, repetition and drop-out data. They are in the form of the number of a cohort of 100 pupils entering grade 1 who will ever reach grade 4. It is thus a measure of the dropping out after 0, 1, 2, and 3 grades. Different countries regularly publish drop-out figures for each grade, and time-series data are often available.

TABLE 7: Weighted average survival rates for cohorts starting primary school about 1980/81

Region	No. of countries	Percentage reaching grade 1	2	3	4
Europe and USSR	19	100	99	99	98
Africa	38	100	83	80	71
Arab States	*6*	*100*	*91*	*91*	*83*
French-speaking	*16*	*100*	*84*	*82*	*74*
Portuguese-speaking	*4*	*100*	*57*	*43*	*26*
English-speaking	*12*	*100*	*82*	*77*	*72*
Latin America and the Caribbean	21	100	76	69	63
Asia and Oceania	30	100	88	81	76
Arab States	*9*	*100*	*94*	*94*	*93*
Other Asia	*16*	*100*	*95*	*94*	*93*
Oceania	*5*	*100*	*95*	*94*	*93*

Source: [33]

Additional information about dropping out can be found from the answers to the questionnaire mentioned above. The extent is considerable in many countries. Colombia reports 62.5 per cent in primary school but 82.4 per cent in rural areas. In Bangladesh there is enormous dropping out: 80 per cent in the whole primary cycle, of whom 60 per cent between grades 1 and 2. In Pakistan and Uganda half of all entrants never complete primary school. Nicaragua presents extremely discouraging figures: only 24 per cent of all entrants in 1972 finished in 1978, and between first and second grade there was a loss of over 50 per cent. Only 6 per cent of primary school pupils in rural areas were ever promoted to the next higher level. In Mexico, the proportion of children dropping out before completing primary school has been decreased to 55 per cent. In Brazil, 50 per cent of all first graders drop out and only 16 per cent of all entrants finish eight grade. Nepal and Senegal report high drop-out figures. Other countries report lower but still considerable drop-out figures, but there are a some countries with little or no dropping out. Among these belong most highly industrialized countries but also countries like Bahrain, Cuba, Cyprus, Malaysia, Nigeria, Qatar, Republic of Korea, Seychelles, Tonga, Turkey and Viet Nam.

Repetition rates and drop-out rates are not stable phenomena. Over the years for which observations have been made and statistics exist, there have been changes. Some of these changes are slow, caused mainly by changes in socio-economic conditions in the countries, others are more abrupt, caused by changes in government policy in respect of education.

Unesco [33] reports average repetition rates for 1970 and 1980. In most countries there are very small changes, often fractions of 1 per cent. Significant drops are reported for Libya (from 26 to 9%), Uganda (from 17 to 10%), Rwanda (from 30 to 6%), Cuba (from 22 to 6%), Iraq (from 21 to 10%), Qatar (from 24 to 13%), Sri Lanka (from 22 to 10%), and Afghanistan (from 27 to 15%). Significant increases are found for Djibouti (from 11 to 21%) and Guinea (from 9 to 22%).

Unesco has also examined sex differences in respect of average repetition rates. They are slightly higher for boys than girls in sixty-six of the ninety countries for which sex-specific data were available in 1980. In 1980, the average repetition rates were generally lower for girls in fifty-two out of seventy countries. The differences are nowhere very great and seem to depend on the fact that the enrolment of boys extends to intellectually and socially weaker strata of the population.

Also in respect of dropping out there is a very slow improvement between 1970 and 1980. In Africa, the proportion of children who reached grade 4 of the primary school increased from 0.73 to 0.79; in Latin America and the Caribbean it increased from 0.70 to 0.73; in Asia and Oceania from 0.86 to 0.91; and in Europe from 0.94 to 0.96. The tendency is not general, however, for in some countries the drop-out rates have increased. A study has also been made of changes in the proportions of children who reached the last grade of primary school in 1970 and 1980. Most countries showed a slight improvement in this case too, but there are quite a few exceptions, for example in Egypt, El Salvador, Kenya, the Libyan Arab Jamahiriya, Panama and Togo. A study was made of sex differences in respect of dropping out. Generally speaking, such differences were extremely small.

There is an intimate relationship between the repetition and drop-out phenomena: a student who fails in school and is forced to repeat a grade is more likely to absent himself from his school and even to drop out than a successful student. Often the choice between leaving school entirely and repeating a grade is determined by other factors than whether schooling is benefical to the child or not. (See below and compare with [4] and [89].)

The effects of repetition and dropping out are immense. It has been shown that a child needs at least six years of schooling to become permanently literate and able to use its literacy skills in daily life [11]. On a world-wide scale, it can be estimated that less than 60 per cent of all enrollees ever reach grade 6 and would thus remain literate; in some countries this is the case with less than 50 per cent. The situation is made even worse by the fact that extremely high drop-out figures are found in some of the most populous countries in the world, for example, Bangladesh, Brazil, India, Pakistan, Thailand and Viet Nam. The cultural effects of a country having a mainly illiterate young population need not be mentioned. There are, however, also other effects: low productivity and unemployment (as pointed out in a questionnaire answer by the Central African Republic, Peru and the United Republic of Tanzania), delinquency (Central African Republic), and a corps of marginal people (Central African Republic). The effects on the child are obviously enormous.

Reasons for repetition and dropping out in school in individual countries have been studied by quite a number of researchers [4], [5], [13], [21], [77] and others; (the research is summed up by Chaïbderraine [4]). There seems to be general agreement among these authors that socio-economic and school factors are the most important ones, the former particularly for determining failure in school, the latter for determining who will drop out and who will repeat. The cultural conflict existing between school and society in many areas, the former often representing a more 'modern' and urban culture than the latter, has been found extremely important.

The Plowden Report [17] and the data obtained by the International Association for the Evaluation of Educational Achievement point in definite directions: life conditions, abilities and attitudes constitute the primary causes for repetition and dropping out. The characteristics of the teacher are of less importance, and factors like the size of the class, the availability of materials, and the physical characteristics of the school explain a small part of all cases. Girod [16] mentions factors like the socio-economic level of the family, the characteristics and behaviour of the child, and conditions in the school, but most of the cases are referred to as caused by 'undetermined factors'.

Many countries answering questionnaires in connection with the thirty-ninth session of the International Conference on Education gave reasons for repetition and dropping out. *Factors in society* were stated by most answers. A dispersed population was mentioned by Chile and Thailand; a severe climate by the latter. The answer from China talks about badly situated schools. The United Republic of Tanzania mentions her migrant population. The cultural conflict between school and people is mentioned by, for example, Brazil, Gabon, Malawi and Zimbabwe. Reluctance towards and disbelief in school is mentioned by Colombia, Poland and Thailand. Jamaica finds that her school system is unable to improve the children's situation, and Colombia reports that the education given is inadequate. A combination of cultural and economic factors is found responsible by Argentina, Jordan and Madagascar. In Zimbabwe there was previously an open-door policy with children flocking to school; this forced the government to introduce levies, and children started to drop out as their parents could not pay the fees.

Among *reasons in the schools* themselves, the following seem important: insufficient facilities and crowded schools (Gabon, Madagascar, Malawi, Senegal, Thailand, Zambia); lack of books and materials and few or inadequately trained teachers (Brazil, Gabon, Madagascar, Malawi, Senegal); an inefficient or inflexible system which does not cater to local needs, or a school programme with inadequate criteria (Argentina, Chile, Gabon, Mexico, Poland, Senegal); incomplete schools (Chile, Madagascar, Zambia); an inadequate examination system (Jamaica); and education conducted in a foreign language (Gabon). In Uganda the population is said to increase more rapidly than the school system can absorb, and in Pakistan many under-age children attend with their older brothers and sisters; when the latter are promoted, the former drop out.

Family poverty is a major reason for wastage, according to many answers. Jamaica mentions that parents cannot always afford to pay costs for transportation, food and materials for their children. Child labour, often at home or in family enterprises, is a major problem according to answers from many countries in Africa, Latin America and the Caribbean. Malnutrition as a cause is stressed by Colombia and Sri Lanka, and physical disabilities by the former. Lack of motivation and negative attitudes among parents are reported by Malawi and Pakistan.

Improvement in the performance of primary education

In the preceding text, a few aspects of the performance of the primary education systems have been scrutinized: student achievement in science, the ability of the systems to enrol pupils, their ability to give equal opportunities to all children, their ability to bring pupils at a normal rate through the system without repetition of grades, and their ability to hold the pupils and bring them to their educational goals. These aspects are important, but obviously they are not the only ones. It is felt, however, that it is possible to draw conclusions from them regarding the performance of systems in general.

Already from what has been shown, it is obvious that the education systems in the world do not always function well, particularly in most developing countries. Countries are aware of the situation and take steps to improve it.

Most countries which have not attained *universal primary education* have plans for doing so and, in the answers to the questionnaires sent out in connection with the thirty-ninth session of the International Conference on Education, various approaches are mentioned. Some of the steps are meant to influence the population. Legal measures are mentioned by the Ukrainian and Byelorussian SSRs to improve the schooling of children. Jordan plans to co-ordinate steps in the schools with action by local authorities so that all children enrol at the age of 6. Peru plans co-ordination with health, feeding, social and other schemes, and Algeria, Argentina, Brazil, Chile, Jamaica and Jordan have introduced or have improved various social actions — primarily feeding schemes. Economic assistance to the needy exists in Argentina and Qatar; Mexico mentions access to educational, cultural, sports and recreational facilities as a means to attract children to primary school. India mentions different strategies to achieve universalization: non-formal and part-time education; linking education with rural development; linking environmental education with real-life situations; and so on. Mexico, Nigeria, Senegal and Turkey place the emphasis on adult education for those not in school. Many countries stress the need for a more even distribution of educational facilities, particularly in poor and rural areas; for example, Jordan, Mauritius, Paraguay, Poland, Thailand and Turkey. Bangladesh, Cameroon, Jordan, Nigeria and Rwanda mention different ways of alerting parents to the importance of education. Argentina mentions the need to set up co-operation with poor groups and sectors of the population; Paraguay desribes its participation in community development; and China specifies local action to end wastage in school. Chile mentions that children with difficulties in normal school shall be rehabilitated so that they can be incorporated into social life.

Improvements in the school system are mentioned. The building of new schools, particularly in rural areas, is found necessary by, among others, Bangladesh, Cameroon, Paraguay, Turkey, Uganda, United Arab Emirates, the United Republic of Tanzania and Zimbabwe. Chile plans new types of schools. The necessity to encourage teachers to go to rural areas is mentioned by Guyana and Nepal. The Byelorussian SSR describes specific measures, including investments in new schools, reduction in two-shift schools, boarding schools, bussing, feeding schemes and so on.

The need to train and retrain teachers adequately, and to improve the quality of education and educational services is mentioned by many countries. New curriculum programmes, particularly in frontier areas and in slums, are mentioned by Peru. Nigeria wants to make schools practical and to teach children a vocation; Malaysia also stresses the acquisition of elementary practical skills. Argentina, Madagascar, Mexico and Paraguay mention decentralization and regionalization. Prolonged schooling is planned by several countries, as has already been mentioned. Concrete steps like distance education, ambulatory schools, independent study and recuperation programmes are also mentioned by many reports. Cameroon wants to update incomplete schools; the United Arab Emirates intends to improve the handling of student flow; and Nigeria proposes to admit Quranic school students into the regular system. Private schools are officially recognized in many places.

Many countries have been searching for a solution to the problem of insufficient economic and other material resources. In some cases, local communities are asked to provide solutions, which mean reduced costs to the State. Shift schools, one-teacher schools and other types of low-cost solutions are mentioned. Literacy courses for young people have been introduced in Sri Lanka and Turkey, and they are planned in Viet Nam. Evening schools exist in the Syrian Arab Republic and schools for gifted young people with accelerated programmes in Nicaragua, Nigeria and Paraguay. A vocational secondary-school programme for drop-outs is planned in Bahrain.

The use of modern media seems common in connection with the spreading of education. This means distance education, radio and TV education, and even a 'Correspondence College' for primary school drop-outs (Malawi). Better book and material production is found necessary for the universalization of primary education by Colombia, Jamaica, Madagascar and the United Arab Emirates, and renewed curricula by Colombia and the United Republic of Tanzania. Bangladesh has created community learning centres, while decentralization programmes to bring schools closer to the people are mentioned by Chile and Tunisia.

Attempts to create *equity of educational opportunities* are often mentioned. Many countries report the problem of children in isolated areas. Australia discussees distance education, boarding schools and the 'school of the air'. Sri Lanka plans to upgrade her small rural schools and train teachers for multi-grade teaching. Algeria, Cameroon and Uganda mention boarding schools for children in isolated or dry areas and for nomads. Nepal pays teachers extra for work in remote areas. Children in slum districts are paid particular attention in Sri Lanka. Working-class families in areas destroyed during the recent war are given priority in Viet Nam, where schools with special curricula have been organized. Orphans in Malta attend special institutions attached to regular schools, and in Colombia and other countries there are programmes for impoverished regions.

Particular attention is paid to minority groups in most countries. In Australia aboriginal schools are provided with aboriginal teaching assistants, who impart knowledge of aboriginal culture and form a bridge with the home. Programmes exist for expanding the number of aboriginal teachers.

Teaching of the national tongue to immigrants and other minority children is provided by many countries. Special teaching of a minority language where this is the mother tongue is found in some countries, too. Australia and Sweden organize events to spread knowledge of the cultures of minority groups, and these countries and Argentina report bilingual and bicultural education.

Other important groups of disadvantaged children are the physically, mentally and socially handicapped. In some developing countries there are few facilities for handicapped children, however, and usually only a small minority of such cases can be sent to special education. Madagascar has a single school for physically handicapped pupils; Malawi, Nicaragua and Rwanda report that they have only a few centres for physically handicapped cases; and Tonga mentions a single special group under the Red Cross. On the other hand, there is general and some vocational education for handicapped children as well as a teacher training centre in Pakistan. The reason for the dearth of facilities for these groups is evidently the high costs associated with such education.

Poland reports special establishments for sick children. Protected workshops or vocational training for handicapped children is mentioned by Cyprus, the Federal Republic of Germany and Turkey. Programmes for treating difficulties with reading, writing and arithmetic are found in many countries, and school psychological services are found not only in wealthy ones but also elsewhere.

Other groups generally given attention are *repeaters and drop-outs*. Many ways of reducing wastage in school have been reported in the answers by different countries. Some refer to social action. Bangladesh, Cameroon, Jordan and Rwanda mention different ways of alerting parents to the importance of education. School feeding schemes and other social programmes have been introduced or strengthened in Algeria, Argentina, Brazil, Chile, Jamaica and Jordan; family allowances exist in Argentina; and material assistance is given by Viet Nam. Administrative measures to return drop-outs are described by the Central African Republic, Jamaica, Jordan, Spain and the United Republic of Tanzania, provision for transport by Argentina and Thailand.

To reduce wastage, Chile describes plans for innovation with new types of schools. Nigeria wants to make education functional and to teach trades, and Malaysia intends to make the curriculum stress fundamental skills. The idea of a new school open to the environment and using the national tongue as the medium of instruction is being introduced in Senegal. Decentralization of administrative services is proposed in Madagascar and Mexico, and a flexible curriculum in, for example, Colombia. In several countries there are schools adapted to local environments and to habits in the communities, and local arts, crafts and languages are mentioned. The provision of more schools, to avoid overcrowding and reduce distances, is mentioned in many replies. In China, some places have made local rules to solve wastage problems.

Some measures to improve instruction in the schools have also been reported: making schools attractive (Mauritius); getting new modern books, guides, equipment and materials (many answers); improving inspection (Spain); training multi-grade teachers (Thailand); improving the evaluation system (Chile); using the new media (Rwanda, Thailand); providing extra teaching (Viet Nam); and so on. Switzerland expects that individualization will enable children to cope with courses. Improvement in the teacher's situation through better training is called for by a large number of countries, and more teachers and better working conditions by Rwanda. Finally, a better direct contact with secondary school is seen as a means of improvement by Cameroon and Nigeria.

Evidently, these measures are essentially the same as those taken or planned to help universalize primary education in the countries where not all children are enrolled in school. A few countries tie the two problems together in their educational plans. It cannot be expected that the steps mentioned above will remedy the situation in all countries in the near future. For this reason other means have been used to achieve some general literacy education. These means include adult education, evening classes for young people and distance education.

Attempts at improving the situation in primary education often meet with *difficulties* and even overt resistance, particularly, it seems, in very poor countries. In most countries, educational renewal and improvement are always restricted by economic and other material constraints. In their work to distribute available resources, responsible politicians find that they cannot provide sufficient funds to education, which often accounts for the largest share of the budget after defence. The economic problems which have been facing the world for about a decade have accentuated the problem.

From 1960, the year for which the first reliable data exist, to about 1975 the education system received a larger and larger share of GNP in most parts of the world. This is shown in Table 8, which has been abstracted from the Unesco publication *A summary statistical review* [36]. Please note that the data refer to public expenditures only, and to expenditures for all levels of education.

From 1975, the share of GNP spent on education in most countries has stopped growing at the same time as the growth of GNP itself has also declined, except in oil-producing nations. Many countries — developed as well as developing ones — therefore mention a lack of resources.

The difficulties facing education systems are not restricted to the economic sphere but to many others as well. The quality and preparation of teachers and other school personnel, as well as their attitudes and values, often form bottlenecks in the renewal process. Organizational and administrative factors sometimes form a hindrance to this effort. External factors, like the geographical and population situation of the country, public attitudes and opinions, laws and agreements, may also play a large part.

Countries have been looking for solutions which allow them to improve their education. In view of the scarcity of resources, it is necessary to utilize existing ones to the maximum. This means not only stating priorities and trying to cut costs but also finding ways of making education more efficient. The solution to the problems mentioned above may be found in a renewal of the education system itself. This will form the theme of Chapter VI of this book, which will also discuss difficulties encountered in this renewal.

made by responsible authorities in different countries. Quotations from original sources have been used quite frequently, however. In all cases, the available material has been official in character, thus expressing authorized views. As these sources expressed different degrees of explicitness, comparison between individual countries is rarely feasible, but a clear picture of the diversity of goals and objectives of primary education emerges.

In order to obtain an idea of the way in which countries express their educational purposes and of the roles played by different educational goals and objectives, it was found useful to develop a model of the latter. This model is based on the assumption that educational purposes can be classified along three dimensions: (a) general *versus* concrete ones, that is, from general goals to concrete objectives; (b) person-centred *versus* society-centred ones, that is, from goals and objectives which have reference to what happens to the individuals in schools to goals and objectives which stress that what happens in the schools shall have a purpose in the larger society; (c) traditional *versus* innovatory ones, that is, from the goals and objectives that serve to maintain the existing situation and have to do with the direct use of what is learned in school to those which are based on the view that the school is a means for development and change.

The model is shown in Figure 3. The dimension general *versus* concrete is expressed as a movement from the center to the periphery. It is assumed that there are goals which are so general or complex that is not possible to locate them along the dimensions; they form a set of core issues. At the level of general goals, however, differentiation is possible.

Traditional society-centred goals are often explicitly ideological or cultural in nature since they imply application of what is learned in school for the benefit of society. Traditional person-centred goals would obviously usually bring up pedagogic issues referring to what should be taught in school. Person-centred goals of change refer to changes taking place in the learner or the teacher: how does the system want this person to develop? Society-centred goals of change imply societal and social changes caused by education.

Concrete expression of ideological or cultural goals leads to religious, cultural, purely ideological and patriotic objectives; only a few can be shown in Figure 3. The educational goals express that the student should, on the one hand, learn for the sake of practical work and further schooling and, on the other, develop products of education such as skills, knowledge, values, experiences, attitudes, etc. The person-centred goals of change are expressed as objectives like learning to think, child development and so on.

The goals of social and societal change find expression in educational objectives for social change, societal change, and cultural and socio-economic improvement. The generalization of education and equity of educational opportunities belong here.

The first part of the present chapter will deal with the goals and objectives of primary education as a whole, without making special reference to particular disciplines. There are, however, goals which do not concern primary education as a whole but only particular areas or subjects. The second part of the chapter will, therefore, turn to the goals and objectives of science and technology education. A general discussion of the findings concludes the chapter.

GOALS AND OBJECTIVES OF PRIMARY EDUCATION AS A WHOLE

This part of Chapter III will be divided into sections corresponding to the parts of the model presented above. First the general goals which cannot be referred to a particular

Goals and objectives 51

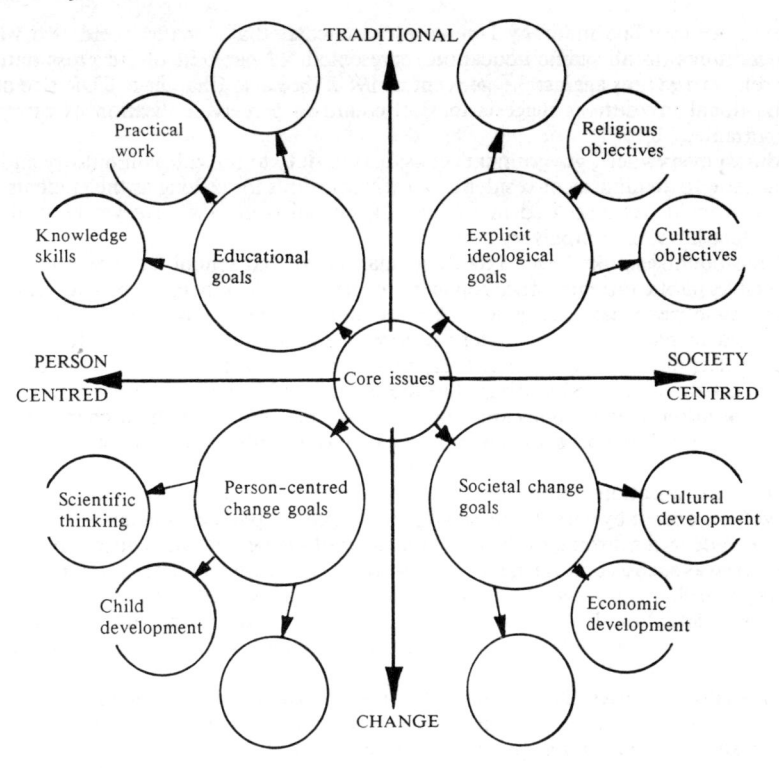

FIGURE 3: A model of educational goals and objectives

quadrant — the core issues — will be discussed. Then each of the quadrants will be treated in turn. Finally, the role of science and technology education seen from the perspective of these goals and objectives will be scrutinized

Core issues in primary education

The Universal Declaration of Human Rights, which was adopted by the General Assembly of the United Nations in 1948, referred to education in the following way:

Everybody has a right to education. Education shall be free, at least in the elementary and fundamental stages. Elementary education shall be compulsory. Technical and professional educational shall be made generally available and higher education shall be equally accessible to all on the basis of merit.

This declaration states two facts of importance to this book: primary education shall be free and compulsory, and education for all shall also include its application in technical and vocational fields. These ideas have now been generally accepted in nearly all countries of the world.

Primary education is felt to be a national task; as a matter of fact, it is one of the most important and economically most demanding duties of national and state governments. An estimate made by Unesco [36] indicates that, for the world as a whole, expenditures on all public education represented 5.7 per cent of the gross national product in 1981, as against 3.6 per cent in 1960. (See also Chapter II.) The size of the educational investment suggests that all countries perceive education as extremely important.

Almost everywhere, governments consider it a duty to provide compulsory and free education to all citizens or residents of their countries for as long as the students may profit from it. As discussed in Chapter II, not all countries have yet been able to introduce laws on compulsory education.

A few countries, like Nigeria and Uganda, report that school fees are paid, but, in principle, public primary education is free everywhere. Yet there are private schools in many countries which charge fees. A few countries report, however, that parents are expected to pay for books, materials, school uniforms or excursions. Rwanda, the United Republic of Tanzania and Zambia mention that parents are involved in the construction of school buildings. On the other hand, many countries mention that transportation, meals, books and materials may be supplied without charge. Poland, Qatar and Saudi Arabia state that material help is provided to poor parents of primary school pupils.

Whether education is compulsory or not and whether it is provided by public or private schools or by parents, accepted political philosophies everywhere stress the role of the State as a guardian of the common good of the people, and education is always perceived as such a common good. What varies are the different perceptions of how best to achieve this common good and, obviously, what the definition of 'good' is in a particular situation. But education is not only 'good' to the people, it is also of importance to the nation itself. Here also there are different opinions and interpretations about what is 'good'.

In socialist countries, the interpretation of what is good to people and the State follows explicit ideological lines. There is often a rejection of past social, societal and political order and an overt commitment to Marxism-Leninism. As a consequence of this, prominent features are the education of the masses on ideological matters, education for loyalty, participation in the development of the country, practical work, and other traits of a socialist nature. Education becomes work related, and the relationship between school and society is stressed. Let us look at some examples.

The Law of the Byelorussian SSR on public education states:

The purpose of public education in the USSR is training educated and thoroughly prepared active participants in the construction of a communist society brought up on the ideas of Marxism-Leninism, in the spirit of respect for Soviet laws and socialist order, communist attitudes to work; healthy, able to work successfully in various branches of economic, social and cultural construction, to take an active part in social and State activities, to preserve and augment the material and spiritual riches, to protect the environment...

Bulgaria mentions the goals of centralization and nation-wide participation, secularity and humane orientation; Poland, those of uniformity of the system, and universal, democratic and lay education; the German Democratic Republic, a unified, public and lay polytechnical education; and Czechoslovakia, a school system which forms an inseparable part of the revolutionary changes in the country and which is unified, democratic and based on principles of equity. The Ukrainian SSR states that the school shall educate people not just to possess certain kinds of knowledge but, above all, as citizens of a socialist society and as active builders of communism.

Viet Nam has a similar set of statements of educational goals, but they are coloured by the fact that the country has accepted communist principles more recently. Cowen [9, p. 19] reports them a follows:

Goals and objectives

1. Marxism-Leninism is the ideology which governs educational progress.
2. To guarantee the rights of the working people is the guiding principle on educational matters.
3. Education comes under the unified administration of the State. The State, in collaboration with the people, supervises the construction and development of socialist education.
4. Education must serve the objectives and political tasks of the socialist State and must be closely linked with everyday life and with measures to construct and defend the socialist nation.
5. Education aims to train future generations of revolutionaries, to train new men, socialists, workers who are aware of the governing role of the collectivity and who have an all-round development...

Ethiopia, which is in approximately the same situation as Viet Nam, states:

There will be an educational programme that will provide free education, step by step, to the broad masses. Such a programme will aim at intensifying the struggle against feudalism, imperialism and bureaucratic capitalism. All necessary measures to eliminate illiteracy will be undertaken. All necessary encouragement will be given for the development of science, technology, the arts and literature. All the necessary effort will be made to free the diversified cultures of imperialist cultural domination and from their own reactionary characteristics...

Madagascar wants education to liberate man and create an understanding of responsibilities for one's duties and one's identity; and Guinea aims to prepare a new man who accepts the Guinean revolution, a man capable of handling scientific as well as practical knowledge who can improve the level of its people and increase productivity, a citizen armed with revolutionary ideology and virtues.

Cuba, to take a final example, states its educational aims in the following terms:

The supreme aim of Cuban education is to achieve integrated and harmonic education of the personality with the view that every individual shall become a conscious and active builder, at the same time as he benefits from the new society that is built. For this reason the national system of education develops, in a unified way, intellectual, scientific-technical, political-ideological, physical, esthetic, polytechnical and vocational, and patriotic-military education.

In the socialist countries there is evidently a feeling of having passed through a revolutionary stage which has changed society and its fundamental principles. These changes must be brought to the people and strengthened, and education is an essential means to this end. In several of these countries, the revolution is a recent phenomenon and is felt as something important that must be defended, as exemplified by Viet Nam. The new system must be strengthened and adapted to the specific conditions in the country, and existing institutions must be adapted to it. Also in a country like Cuba, where a socialist ideology has been accepted for a whole generation, there is a feeling that a new society is being built upon a new foundation.

In other cases, the basis for the development of the education system is in the form of religious principles. In such situations, countries experience the rapid changes taking place in the surrounding world and try to reconcile the need for educational and socio-economic development with ideological-religious principles. These principles mean that a framework is imposed on the education system, but at the same time the system itself and renewal efforts within it are legitimized and strengthened.

Most of the countries which state that they base their education systems entirely or largely on religious principles are Moslem ones, but religious thought is evident in other cases, even in secularized countries in the Western World.

Saudi Arabia, to take an example of a Moslem country, states the fundamental principles and main objective of education in religious terms. The following points are taken from *Educational policy in the Kingdom of Saudi Arabia* [24]:

A. *General principles of education:*
1. To believe in Allah as the only God; Islam as the only religion and Mohammad (May peace be upon him) as God's apostle and messenger.
2. The total Islamic concept of life and universe, that the entire world is subject to the laws of God in order that each creature would fulfill his duty without interruption or confusion.
3. Life on earth is a stage of work and production during which the Moslem invests his capacities with a full understanding of and faith in the eternal life in the other world. Today is work without judgement, and tomorrow is judgement without work.
4. The message of the prophet Mohammad (May peace be upon him) is the soundest for leading a virtuous life and it ensures happiness to man and rescues humanity from all corruption and misery...

B. *Purposes and general objectives of education*
1. The purpose of education is to have the student understand Islam in a correct and comprehensive manner; to plant and spread the Islamic creed; to furnish the student with the values, teachings and ideals of Islam; to equip him with the various skills and knowledge; to develop his conduct in constructive directions; to develop the society economically, socially and culturally; and to prepare the individual to become a useful member in the building of his community...

Basing education on religious principles and teaching religion in school does not mean that Saudi Arabia neglects other aspects. Other objectives of education, such as those concerning modern science, are derived from religious ideas.

Norway, a fairly secularized European country, emphasizes the role of religion in explaining cultural identity and development in a modern democratic state. For this reason, a Christian-moral education is found necessary:

The school must help to pass on the cultural heritage of Christianity to the coming generation. The basic values stressed by the Plan must, therefore, have a central position on the curriculum [9, p. 23].

A large number of other countries see religious ideas as a basis for values in society, like honesty, loyalty, solidarity, respect for others, and humanity. A positive religious upbringing intended to make children believers in a faith is also reported by a few countries. It is, however, primarily expected that religious teaching shall lead to personal qualities useful to society and valuable in the individual. Whether the religion in question is Islam, Christianity, Hinduism or something else, it is strongly anchored in its social context and it serves to strengthen social behaviour. It thus aims at bringing about an awareness of the necessity to work for the future of this society.

It would be incorrect to assume that an ideological and religious foundation of education is found only in countries whose education is characterized by explicit ideological or religious goals. On the contrary, both an ideological and a religious basis is characteristic of all education systems with which the present authors are familiar. To take an example: Törnvall [80] showed that the general basic view behind the Swedish curriculum is characterized primarily by a secular humanistic outlook coupled to a socialistic outlook, while the view of the individual found in the curriculum shows a secular humanistic basis for the development of the pupil's personality towards self-realization in collective fellowship. In other countries, the ideological basis may be political conservatism, for example, in the United States, or Christian humanism, as in many other Christian countries. It is also a well-known fact that Western thought and Western culture still depend heavily on Christian ways of thinking, and this penetrates deeply into the given goals and objectives of education. The ideological and religious basis for the education given is not always explicitly stated, however.

Cowen [9] reports a third principle on which education systems are built — the national constitution. Although evidently not of the same order as a political or religious ideology, a constitution still furnishes an essential guiding principle for the development of an education system. It has been shown that behind a constitution may lie an ideological or religious system. A few examples illustrate the diversity of educational goals:

In Greece, the governmental policy rests on the following considerations, as expressed by the Prime Minister:

Education is a public good, and every citizen has a right to it. The obligation of the State to guarantee this right of young people is an immediate priority in all parts of our country. This means a schooling which is identical, of constant quality, with coherent and obvious procedures; objective and based on merits; an information which is complete and credible about possibilities and conditions for studies at each level, about the concrete needs of the economy and production and about employment conditions. One of the fundamental objectives of education is to detect and develop creative possibilities and talents in our young people. It is also necessary to prepare them systematically to undertake, as responsible, conscientious and critical workers and citizens armed with satisfactory scientific and technological knowledge, the difficult work of renewing and developing the country.

Venezuela wants to undertake the necessary educational reforms so that education assumes the characteristics, aims and cultural values of each village. Bahamas states that the education system 'shall contribute towards the spiritual, moral, mental and physical development of the community by ensuring that efficient education throughout these stages shall be available to meet the needs of the population'. In Mexico, the education law states, among other things, that education shall take care of the harmonious development of all human faculties and, at the same time, foster patriotism and awareness of international solidarity, independence and justice. This law also stresses that education is the foremost means to acquire, spread and improve culture; that it is a permanent process which contributes to the development of the individual and the transformation of society; and that it is a determining factor in the acquisition of knowledge and for the creation of a person who is conscious of social solidarity.

Morocco mentions the generalization and democratization of primary education, the Moroccanization of teaching staff, and the Arabization of teaching; the Syrian Arab Republic states that the educational and cultural system aims to create a generation of nationalistic, socialistic Arabs with a scientific mind, attached to their history and soil, proud of their country, full of a fighting spirit to realize the aims of the country in unity, freedom and socialism, and ready to participate in the service of human progress. Kuwait mentions the right of each citizen to have as much education as possible according to his abilities and potentials. In these Arab countries, Islamic thought is also very influential in determining educational goals.

Independently of the ideological and constitutional basis for education, it is subject to considerable pressure from the immediate surroundings. Changes in the larger system constituted by society have an effect on the subordinate education system. The rapid societal and social changes taking place in most countries, as well as the rapid growth in science and technology, affect education at an accelerating rate. This has direct consequences for the inclusion of new elements into what is being taught in school, for example, information about computers and modern communications. It may also have an effect on the organization of education, such as when new political ideas demand decentralization or when it is intended to adapt education to local conditions, or on teaching methods when new materials or other instructional resources become available. The effects may also be indirect. New priorities occur in society and modify the education system, new learning styles and new motivational factors influence work in school, and the relationship between school and community life may change.

The recent scientific and technological advances pose a serious challenge since a country will be left behind in the race for development unless the education system can cope with the new tasks. Socialist countries seem to be aware of this problem, and their educational goals emphasize science and technology, the relationship between education and society, and practical work experience. Other countries feel the need for renewal. In the United States, the National Commission on Excellence in Education has produced a report, *A nation at risk* [50], which stresses the need for educational reform.

Most countries have adopted recent legislation, plans or programmes which aim to bring educational development up to a modern level. As examples, one could mention the five-year educational plan 1980-84 in Algeria; the development plan for the educational sector 1983-86 in Colombia; the five-year educational plan 1982/83-1986/87 in Egypt; the ten-year indicative plan beginning in 1984/85 in Ethiopia; the educational plan in Iraq; the national programme for the development of education, sport and recreation 1984-88 in Mexico; the national medium-term educational development plan for 1983/84 onwards in Nicaragua; the 1979 national educational policy and implementation programme in Pakistan; and the transitional national development plan in Zimbabwe.

The issues and goals described above are extremely general. It is hard to make a clear distinction between such general, undifferentiated purposes and educational goals of a more specific nature, since the demarcation line is quite arbitrary; it is equally hard to differentiate between general goals and concrete objectives. This problem has no real consequences, however, except that a reader might find that related facts are reported in several places.

Explicit ideological, cultural, and political goals and objectives

The first main group of goals and objectives should, according to the model, be characterized by being society-centred and traditional. Learning because society calls for it might be another way of stating their common feature. It is not learning because the individuals need it; such learning is covered by the goals and objectives placed in the second quadrant of Figure 3. It is not learning because we want to change society; this belongs in the fourth quadrant.

The types of goals and objectives which belong in the first quadrant are obviously those preserving something valuable existing in society: its ideology, its culture, its organization, its religious affiliation, its strength, and so on. The goals and objectives which have been mentioned by countries are, however, mainly ideological (including religious), cultural and political ones.

Goals. As mentioned, there are countries which explicitly base their whole educational policy on religious ideology or feel that religious education and a religious upbringing are important goals to be pursued. There are also countries which find that religious ideas are important to an understanding of culture.

Qatar is an Islamic country, and positive religious education is considered essential. The cultural heritage is also stressed, as can be seen from the following statement [9, p. 46]:

Goals and objectives

Through inculcating faith in the oneness of Allah almighty, thus creating a feeling of piety and obedience of all ordinances revealed by Allah; and: Through the application of Islamic ideals and values in everyday life as a direct impact of including such values in the content of education... Preservation of the country's heritage by identifying the impact of Islamic Legacy on all fields of knowledge, thus identifying the pioneering role of Islam and its contribution to human civilization.

Pakistan is another Islamic country where religious thoughts penetrate into State policy and everyday life. Priority is given there to the revision of school curricula according to the Islamic faith in such a way that this faith penetrates into the teaching of young people: pupils shall be acquainted with the fundamental tenets of their faith and motivated to practice it.

In some countries religious education is neutral. Sweden is an example. Religion shall be taught in school within the framework of an integrated subject. This education shall be confessionally neutral, however: different faiths are discussed and their relationships and specific characteristics are studied, but the pupil is left to discuss his personal religious choices with his parents, representatives or churches, or others. In this respect Sweden differs from, for example: Norway, where Christian faith is given a prominent place; the United States, where the constitution does not allow religious education in public schools but refers it to churches and religious groupings; and the socialist countries, which are expressedly secularized and unwilling to encourage religious education.

As well as religious ideals, political and cultural goals include those of patriotism, national language, strengthening the ideological foundation of the country, linking science and technology with culture, independent thinking, and so on. To a very large extent, the stress is placed on the integration of various aspects of education. A few examples of goal statements may show this: in Hungary the school should work for the commitment of the future generation towards socialism and progress; the young person should not be just a spectator of social, economic and political events but should be involved. Schools shall educate in such a way that students learn to respect the values of socialist society and so that patriotism is strengthened.

The Byelorussian SSR mentions the unity of education and a communist upbringing, the relationship of school with life and practical tasks of communist construction.

Madagascar stresses the application of what has been learned in school to contemporary life in accordance with socialist ideology. The United Republic of Tanzania states that education shall foster social goals of living together for the common good; schools shall impart socialist values, attitudes and knowledge which will enable the pupil to play a dynamic and constructive part in the development of society, inculcate a sense of commitment to the total community, and help pupils accept values appropriate to the future.

Bolivia describes education as revolutionary and based on a new doctrinal content in the perspective of transforming the spiritual orientation of people; it is anti-imperialist and anti-feudalist; and it aims to achieve the economic emancipation of the nation.

An educational goal in Kenya is to foster national unity based on the adaptation of the rich cultural heritage of her people; Gabon wants education to safeguard cultural identity; Senegal aims to train people who respect positive traditions; and Nigeria aims to inculcate national consciousness as well as correct values and attitudes.

Chile strives to ensure the agreement of education with Christian-humanistic principles. The training shall enable people to live as responsible citizens in accordance with accepted values and habits.

In Kuwait an ultimate goal of education is to establish a progressive society which adopts continuous learning and is able to face the challenges and demands of this age and contribute to the development of Arab and human civilization. Jordan wants the preparation of the individual for responsible citizenship, the development of the individual to emphasize the morals of his social behaviour, and the development of respect for work and democracy in dealing with others. Tunisia wishes to give students a national culture; Bahrain to prepare them to become devout Muslims, patriots, well-educated people, responsible citizens and active members of society.

Objectives. From the general goals, which have been termed ideological, cultural and political in the model, the different countries have inferred concrete objectives. These can be of many kind since they shall express ways in which society can be served by the school system. The material allows statements about religious, cultural and political objectives.

It is to be expected that countries which build their education systems on a religious basis, also mention *religious objectives*. This is the case with Pakistan, which states that an objective is to acquaint the pupil with the fundamental tenets of their faith and to motivate them to practice it. Chile and Colombia, amongst other countries, mention Christian values. Senegal wants the education system to integrate religious values in Senegalese society. Qatar and other Muslim countries aim to bring up children on a sound Islamic basis, and in Egypt religious education of some kind is compulsory.

Saudi Arabia presents a long series of detailed religious objectives; the following examples may show the way in which they are expressed:

- Preaching the Book of God (Quran) and the Sunna of His Prophet (May peace be upon him) by safeguarding them, abiding by their teaching and acting in compliance with their commands.
- Providing the individual with necessary ideas, feelings and powers which will enable him to carry the message of Islam. Enforcing Quranic morality in the Moslims and emphasizing moral restraints for the use of knowledge.

Iraq, finally, wants to create a new generation fully believing in God and to consolidate faith 'in God and his divine message'.

Religious objectives are obviously closely connected with cultural ones. There are, however, countries which express *cultural objectives* without referring them to a religious context.

The Byelorussian SSR aims to make children conscious builders of communism with a scientific outlook and communist morality. Czechoslovakia aims for a materialistic outlook, and Romania wants to develop a collective feeling and a desire to help one another, love of justice, modesty, social responsibility, and respect for laws and social norms.

India mentions the development of social and civic values. Rwanda finds it essential to integrate pupils into their environment. Chile stresses values like patriotism, love of family, respect for Man, tolerance and compassion in the educational programme; and Paraguay wants children to be familiar with and appreciate the primary manifestations of national culture, respect the great persons in the country, and develop patriotism. Children should also acquire habits and attitudes of social value like responsibility, companionship, respect and co-operation.

Mexico mentions a series of cultural objectives: to integrate the family, school and society; to assimilate, enrich and transmit one's culture at the same time as respecting other cultural manifestations; to combat ignorance and all types of injustice, dogmatism and prejudice. Guyana states the objective of helping each child to integrate with others in socially accepted ways.

Saudi Arabia states a series of cultural objectives, of which the following may be quoted as typical examples:

- Providing the student with the necessary information and various skills which enable him to be an active member of society.
- Strengthening the student's feelings about the cultural, economic and social problems of his society and preparing him to combat those problems.

Iraq also states a number of such objectives, among others:
- Establish the principle of democracy as the focal point of a well-organized life system based on freedom, equality and rendering equal opportunities to all.
- Develop will-power in man's personality and his ability for construction and active participation in social solidarity.
- Holding a cultural standpoint characterized by originality, which implies preservation of the national identity, on the one hand, and innovation which implies alteration of the present to a better future through development and progress, on the other.

The last set of objectives to be mentioned here are those which could be called political objectives: objectives of national pride, internationalism, patriotism, work preparation and so on.

Romania wants to develop patriotism and love of the communist party and the Romanian people, and aims to teach students to accept sacrifices to defend what has been gained by the people since the independence of the nation. The country also strives to make workers experience a feeling of unity and brotherhood. At the international level, Romania wants her young people to feel respect for other nations and solidarity with young people in all countries, as well as with democratic forces all over the world which fight for peace, liberty, justice and social progress.

In Hungary, an objective is to train pupils for discipline in work and to make them realize that socialism calls for a working society: regular, disciplined, industrious. The Byelorussian SSR calls for the inculcation of Soviet patriotism and socialist internationalism.

Thailand wants the school to produce law-abiding and good citizens. An objective in the Republic of Korea is to enable pupils to support themselves in life, to fulfill their public duties, to contribute to the development of a democratic state, and to help attain the ideal of common prosperity.

Zambia states that the intention is to give the pupils the basic skills, knowledge and attitudes needed in order to realize their potential as individuals and to become effective participants in the advancement of the country. In Chile an objective is to ensure that school programmes support national reality. These programmes shall be flexible so that they can vary according to needs.

Mexico wants the student to understand the present situation in the country as a result of various national and international factors, to know and appreciate national values and show patriotism, and to develop a feeling of national and international solidarity based on the equal rights of all human beings and all nations.

Guyana wants to develop in each child a feeling of patriotism, and to maintain communication between home and community to meet the needs of children. Iraq stresses the awareness of the national and humanitarian message in the pupil and loyalty to country and nation. The Syrian Arab Republic wants to educate the Arab citizen to love his country, the liberation of the Arab nation, and the attainment of the goals of his Arab homeland. This education should also strengthen socialist values on a scientific basis, to stimulate the young person for the liberation of the Arab nation and for the liberation of Arab society from exploitation, and to prepare the citizen for a role in democratic popular institutions.

Saudi Arabia aims to acquaint the student with the great Islamic glory of his country, its deep-rooted civilization, its geographic, natural and economic characteristics, and its important position among the nations of the world. Qatar, to wind up this list of examples, presents the objective of inculcating in the minds of the children a feeling of national pride and an affiliation to their homeland and its values, traditions and heritage.

Educational goals and objectives

The second quadrant in Figure 3 represents traditional, person-centred goals and objectives. These answer such questions as to what the individuals or the school system as such can get from the learning situation. They can be expressed in the form of expected behaviour on the part of the pupil (or sometimes other person); in the form of educational outputs like knowledge, skills, attitudes, values, etc.; or in the form of a learning situation created in the school with curriculum elements, methods of teaching, means used in the teaching situation, and so on. (See [91], p. 116 et seq.) All this is covered by the heading of educational goals and objectives.

To many educators this is the very centre of the educational debate; and there is a long experience of stating such goals and objectives in most countries. This has resulted in a rich variety of answers by the different countries, and more than in any other field, detailed and concrete information is given.

Educational goals. A wide variety of curriculum areas are mentioned. Since the thirty-ninth session of the International Conference on Education stressed science and technology in primary education, it is to be expected that science and technology goals would be mentioned in a great many of the questionnaire answers. Surprisingly, this is not really the case. Science and technology goals are only mentioned in passing, and few countries place any emphasis on them. On the other hand, many countries stress practical activities and training for work. This is the case with the polytechnical school system in socialist countries, where a main goal is to train students for future work, but it also occurs in other countries.

A few countries mention curriculum changes as a general goal without specifying particular areas. Romania wants to place the accent, within the framework of a system of multifaceted training, on scientific subjects, general cultural subjects, socio-political subjects, specialized subjects, and the acquisition of practical skills. The United Kingdom mentions programmes to change curricula and examinations, particularly to improve mathematics and science education; Ireland keeps curricula under continuous review; and Luxembourg mentions a reform of content and method. In Pakistan and Japan there is a curriculum improvement with the emphasis on simplification.

Czechoslovakia describes curriculum content based on a Marxist-Leninist world outlook which is scientifically modern. The emphasis is placed on progressive cultural traditions, while part of the content aims to develop an all-round personality. The Soviet republics mention the scientific character of their curricula, their constant improvement on the basis of the latest achievements in science, technology and culture, and their humanitarian and moral character.

In Bolivia, education is scientific, based on bio-physical knowledge of the student and providing a systematic training based on scientific processes; it is related to national reality. In Kuwait there is a development of content, methods and techniques of education of various types to create a better development of self-education skills through the application of modern educational technology, and in Bahrain there is a curriculum development to meet the needs of both the individual and the community.

The United Republic of Tanzania gives priority to science and technology education in response to the country's call for the development of physical and human resources in order to create basic industries.

A few examples of goals related to practical work may show the trend; as mentioned, it is intimately related to polytechnical ideas. Romania wants to provide an education with a wide profile to her students so that they will later be able to integrate rapidly into production and social life, and so that, if the need appears, they would be able to handle different types of work within the same field. Students should also participate in productive work. Bulgaria mentions the continuity between structural components of the education system and a direct link with life; and the Ukrainian SSR combines study with productive work, fosters in children a love of work and the habit of doing productive socially useful work, and relates education with life.

India has increased the vocational basis in her curriculum, sought more relevance with socio-economic needs, and developed alternative strategies; there is also an awareness of environmental problems. Bolivia states that education will become active, vital and oriented to work; and that it will give the learner practical training for productive, socially useful activities.

Kuwait recognizes the value of work in personality development and the importance of having good relations between educational institutions on the one hand, and the public sector and private establishments on the other. Breaking the traditional barrier between academic institutions and technical-vocational institutions and adopting new forms of public education is also a goal.

Educational objectives. Objectives should normally be derived from general goals, although this is not always the case. The objectives found in the second quadrant of Figure 3 should therefore refer to the way in which teaching in school should be organized: curriculum content; teaching methods; the organization of the educational and teaching-learning situation; the use of various materials, equipment, textbooks and other teaching aids; the provision of buildings; and the organization of the timetable; and so on. Many of these curriculum and similar elements have been mentioned, but only those of curriculum content, practical work as part of schooling, teaching methods, and the relationship between education and social life have been brought up by more than one country. They are, however, important aspects from the point of view of science and technology teaching in primary school.

For obvious reasons, many countries state the objective of providing *basic reading, writing and arithmetic skills,* for example, Cameroon, Central African Republic, China, Colombia, Congo, Guyana, India, Jordan, Malaysia, Nigeria, Pakistan, Qatar, Rwanda, Saudi Arabia, Seychelles, Thailand, Uganda, the United Republic of Tanzania and the United States. It can be taken for granted that this objective is accepted in other countries.

Many countries have preferred other ways of expressing their needs for basic skills. In an answer from the United Kingdom, it is stated that education shall extend children's knowledge of themselves and of the world in which they live, and develop skills and concepts through greater knowledge. The Netherlands expresses the objectives of familiarizing the child with certain cultural skills (reading, writing, arithmetic and language) and providing an initial understanding of the surrounding cultural, social and natural environment.

Bulgaria wants the child to master general (polytechnical) education with a broad profile. Hungary states that schools should enable students to acquire knowledge and that they should meet contemporary needs. Czechoslovakia expects to provide better quality training so as to enable students to participate in social and public activities.

In Pakistan the basic skills shall enable the child to learn and speak. China wants the child to acquire knowledge of nature and society, and good habits of study. In Turkey the objective is the child's acquisition of knowledge, *savoir-faire* and behaviour indispensable to a good citizen; education shall be in conformity with national morals. Thailand wants permanent literacy and numeracy, and sufficient skills and knowledge to enable people to earn a living in keeping with their age and capabilities. Malaysia stresses, besides literacy and numeracy skills, values and attitudes.

Kenya expresses the objective of providing skills in the following way:

> Education must prepare and equip the youth of the country with the knowledge, skills, and expertise necessary to enable them collectively to play an effective role in the life of the nation and to enable them to engage in activities that enhance the qualities of life.

Gabon states that the child should learn the value of knowledge and intellectual skills by applying them. Senegal wants to train open-minded and informed people capable of adapting to necessary innovations in society and of making pertinent changes. In Nigeria, the aims include permanent literacy and numeracy and effective communication, health and physical education, moral and religious education, and the encouragement of aesthetic and creative activities.

In Chile, an educational objective is that the individual shall be enabled to reach a reasonable degree of personal development. He shall also be made to understand reality in its personal, social, natural and spiritual dimensions. Mexico wants children to be able to communicate their thoughts and their feelings. Colombia aims to give the learner not only certain information but also basic concepts, organizational principles and methodological elements which allow him to plan problem solving correctly and find adequate solutions. The school system shall offer experiences and training in a systematic and interdisciplinary way which facilitate and integrate the interpretation of reality.

Jordan stresses the development of the individual's intellectual abilities and basic skills in several fields. Iraq aims for the acquisition of the tools of basic knowledge and a basis of Arabic and Islamic culture as well as useful citizenship, and wants to foster love of science and a desire to continue learning. Saudi Arabia wishes to provide the student with the necessary information and skills which enable him to be an active member of society. Bahrain states the objectives of providing each individual with the opportunity to develop his/her abilities, skills and attitudes, as well as the socio-economic and cultural level. Qatar wants to provide children with basic instruments of knowledge, such as reading, writing and arithmetic, and to train them to use such means properly. It is also required that children should be made familiar with the necessity of having a clean body, clothes and environment; they should also understand their local environment and become acquainted with the existing sources of wealth and fields of activity.

Besides religion, which was discussed above, and basic skills, few disciplines have in fact been mentioned in the lists of objectives stated by countries. This may be due partly to the fact that not many countries express their objectives in sufficiently specific terms, and partly to the fact that systems of integration of subjects are generally accepted. (See Chapter IV below.) National languages are mentioned by Senegal, Burundi and some Arab states; socialist ideology by socialist countries; and environmental science by several countries.

Few countries mention the need for the teaching of *science and technology*. In most of these cases there is a clear bias in favour of science over technology; the latter subject is hardly ever mentioned.

In the answer from the United Kingdom the need for more time for subjects like sciences is emphasized. Spain mentions the acquisitions of knowledge about nature and the social environment.

All socialist countries that have introduced polytechnical education stress that this education is scientifically based and integrates theoretical, humanistic and science learning in school with practical and pre-vocational experiences in society.

Goals and objectives 63

China mentions the need for knowledge about nature and society. Cameroon wants children to be alert to realities in their rural environment. The United Republic of Tanzania finds it important to give priorities to science and technology in response to the country's call for the development of resources. Congo aims for an initiation to the study of nature and other disciplines. In Ethiopia, encouragement will be given to the development of science, technology, arts and literature.

Mexico wants children to be able to contribute actively to the maintenance of an ecological balance. Jordan mentions natural and social sciences; Iraq, an interest in science. Qatar aims to help children understand their local environment. Bahrain wants to help people take full advantage of advanced science and technology without becoming enslaved by them.

Some answers emphasize the objective of a *link between what is taught in school and what is happening in society.* This is the case with the socialist countries as a group, as was mentioned above.

Bahrain states the objective of ensuring economic and social progress by providing sufficiently skilled manpower. In the United Republic of Tanzania, education is provided to enhance the people's quality of life through the upgrading of skills and knowledge as a means of furthering economic and social transformation. Turkey mentions education for active life preparation. The Republic of Korea states the objective that education shall make the learner able to support himself in life. Uganda aims at a needs-based education which produces people who can live productively in society.

Leisure-time activities are mentioned by Bahrain and Qatar, which state that education shall help people benefit from their leisure time by introducing them to new hobbies and providing recreational facilities.

Primary education as a *preparation for practical work* is a theme in some of the answers.

Spain mentions the objective of making students acquire knowledge, skills and habits which facilitate vocational and professional choice. Bulgaria aims at correct vocational education to enable the selection of professions in accordance with interests, aptitudes and capacities, and with the economic and manpower requirements of the country. Hungary states that the schools should prepare students to join in with the social division of labour; that technical-vocational education is increasingly important; that vocational training should start at an early age, the foundation for which shall be prepared by the school; and that the schools shall provide pupils with general basic knowledge flexible enough both to serve as a basis for further attainment and to enable young people to leave school and adapt themselves to labour regulations.

Gabon wants to inculcate in each pupil a taste for work, while vocational education should be better adapted to actual economic requirements. Malawi aims to enable people to earn a living in a rural setting. Nigeria mentions local crafts, domestic science and agriculture. Zimbabwe wants to integrate education with production and endeavours to prepare students for the world of work.

Chile aims to adapt technical-vocational education to the needs of the nation and gives preparation for work as one of the objectives of primary education, besides general education and preparation for further studies. Tunisia wants to train pupils for active life.

Paraguay has adopted the objective of creating positive attitudes towards work as a condition for self-realization. Jordan wants to develop positive attitudes and respect for work. Iraq wants to instill in the minds of young people the love for manual and practical work, to teach them correct methods of work, to familiarize them with its implementation, particularly in agriculture and industry, and to train them to use simple tools. The Syrian Arab Republic describes the aim to make people aware of the nobility of work and to glorify productive collective work in order to create a society without exploitation.

The objective of making the student a *productive member of society*, mentioned by several countries, is expressed in different ways.

Rwanda wants to make the pupil productive. Egypt states that basic education should create a close relationship between schools and productive work through pre-vocational education and practical training included in the timetable. India has introduced a vocational bias in her curriculum, which has become more relevant to socio-economic needs and developed alternative strategies. Denmark also states that the development toward fuller provision of education means not only prolonged education but also training for a job.

Practical work as part of the curriculum is a fairly common phenomenon and is sometimes mentioned among the educational objectives. Romania states this in the following terms:

...let teachers and pupils (students) take part in productive work, technical and science research, cultural and artistic movement, and the realization of economic and social objectives in Romania.

Chile also wants a combination of study and work in the educational process to prepare the pupil for social life. Bolivia states that education shall be active, vital and oriented to work, and that it should give the learner a practical training for productive, socially useful activities.

Kuwait recognizes the value of work in personality development and wants to break the traditional barrier between academic institutions and technical-vocational institutions, as well as adopting new forms of public education.

Teaching and learning methods are sometimes mentioned among the educational objectives, although mostly in very general terms.

Australia wants to take the differing talents, interests and needs of individuals into account, and India wants a curriculum that is both specific and flexible. Gabon aims to integrate traditional artistic education in school, to develop national art, and use traditional theatre, thus safeguarding its national culture. Mexico states that pupils shall participate in group work in an organized and co-operative way.

Colombia mentions as an objective the integrated development of theory and practice, wherever possible, with the view that the learner, who is familiar with the necessities, problems and real possibilities, shall develop his critical thinking, creativity and willingness to take part in social transformation. Saudi Arabia, which makes Islam the foundation for education, states the following:

Studying the great and wonderful things in the large universe and discover the secrets of the Creator, to profit therefrom and put them in the service of Islam and the dignity of the Islamic nation. — Demonstrate complete harmony between science and religion in the Islamic law.

In the Federal Republic of Germany, it is intended that pupils attending the *Grundschule* (primary school) shall evolve gradually from play-oriented forms of learning to systematic learning. The United Republic of Tanzania wants to stimulate creativity within a context which emphasizes science and technology. Nigeria, finally, stresses the training of the mind to understand the surrounding world.

Goals and objectives of person-centred change

The lower part of Figure 3 refers to systematic development processes, development which may change something in the school system itself or may use what is going on in

Goals and objectives

this system to create a change in society. Person-centred change processes have always been extremely important educational goals. Education has sometimes been defined as a process of systematically inducing change. By means of the teaching-learning process in school, an attempt is made to change the characteristics of the learner in a desired way. However, the change process may also extend to teachers and other individuals, groups, relationships between individuals and groups, and so on. Educational goals and objectives may be expressed in these terms.

Goals of person-centred change

Goals of person-centred change can be of many kinds: child development in a certain direction; the development of abilities to solve problems and think in new ways; changes in teaching and approaches in the classroom; and so on. The goals which have been mentioned have mostly had to do with child development in some form. As is easily understood, it is extremely difficult to distinguish between educational goals involving, for example, child development and educational objectives of this nature: the difficulty of expressing such objectives in concrete terms is enormous, and in most cases the statements that appear are in very general terms, even when they are meant as guidelines to teachers, administrators, or authors of textbooks.

The most common goal of child development is the development of a multi-faceted, 'rich' personality, often expressed as the development of the whole child. This goal is stated by countries in all parts of the world; as a matter of fact, very few countries do not include it in one form or another. A few examples may illustrate how it is provided.

France expresses the goals of developing intelligence and artistic feelings, as well as manual, physical and sporting abilities; education shall be moral and social and is undertaken together with the family. Ireland wants the child to be enabled to live a full life.

Bulgaria states the goal of educating universally trained individuals, citizens with active positions in life; of preparing comprehensive and efficient individuals; and of enabling social and cultural realization. This means stimulating intellectual, moral, physical and cultural development of the student and promoting aesthetic education as a means for multilateral and harmonious development.

In Hungary, schools are intended to bring about the all-round development of personal abilities and faculties. Czechoslovakia wants to educate all-round and harmoniously developed builders of socialist society, people with communist attitudes to work, who are able to participate actively in social and political activities. Poland mentions the goal of primary education in creating a multi-faceted personality in the pupil, of preparing for contemporary life and for work in several fields of human activity, and of shaping a person who wants to become aware of social, natural, cultural and technical realities. In the Ukrainian SSR, the goal is the all-round development of the individual, of his spiritual and moral attitudes, the upbringing of the young generation in the spirit of communist ideals; and the unity of ideological, labour and moral education.

China wants the pupil to achieve full development, both physically and morally, with strong physique and good living and working habits. Tonga talks about the development of the whole child as an individual and as a member of Tongan society. In Malaysia, the goal is the total and balanced development of the child; the intellectual, spiritual, physical and traditional development, as well as the development of the pupil's talents and the inculcation of moral, aesthetic and social values. Kenya stresses opportunities for the full development of individual talents and personality. Gabon wants a multi-faceted education which contributes to an integrated development in moral, civic, physical, intellectual, artistic and vocational respects.

The goal in Mexico is to let the child develop physically, intellectually and emotionally in a healthy way. Colombia emphasizes the need to respect the cognitive aspects of development as well as the social, affective and psycho-motor sides. A goal in Peru is to develop cognitive, will-power and physical faculties considered to be the base for integrated learning, to encourage creativity, to orient vocational development and to stimulate habits related to security, orderliness, hygiene, etc.

In Jordan a goal is to develop the pupil physically, intellectually, socially, emotionally and spiritually, and to let him discover his own attitudes, abilities and capacities. Egypt mentions the basic educational aims of helping children achieve integrated and well-balanced development, equipping them with the basis for being alert and productive citizens, with values of religion, good conduct and patriotism, and with knowledge, attitudes and practical experience. In the Syrian Arab Republic, to take a last example, the goal is a balanced integrated development of the pupil on the physical, psychological, social, moral, national and affective levels by giving such concepts and attitudes that he can find his own way in life.

Other goals referring to child development have been mentioned less frequently. The ability to think and solve problems has been stressed by a few countries, for example, Guyana, Mexico, Paraguay, the United Kingdom and the United Republic of Tanzania, but a number of countries express objectives referring to specific skills of thinking or problem solving. Primary education which provides basic tools for further education is demanded by Nigeria, and it is obvious that this is a basic goal for first-level education everywhere. A needs-based education enabling people to live productively in society is an expressed goal in Uganda, but it goes without saying that this is also a goal behind all primary education.

In connection with the discussion of ideological goals, the needs for religious and ideological training were mentioned. The discussion dealt with these needs from the point of view of society. There are, however, countries which report such goals as seen from the point of view of the individual. Hungary wants scientific education for a personal ideology (*Weltanschauung*); China wants to educate children to love socialism; and Pakistan, Iraq, Saudi Arabia and many other countries want their children to develop a personal religious conviction.

Objectives of person-centred change

It could be assumed that many very different objectives would be classified under this heading; the development of the pupil, the teacher and other people related to the school; the development of the group or collective in the school or of relationships between people in the school; the development of the school as an institution; and so on. Looking at the answers to the questionnaires and at other documents available to the present authors, it is found that the major part of the expressed objectives which can be classified into the third quadrant are those related to the individual learner. Furthermore, most statements refer to his intellectual, physical, aesthetic, moral and social development.

This impression of a myopic look upon the objectives is not entirely correct. While the statements found in documents stating educational objectives refer almost exclusively to the pupil, there are other statements, for example, those in plans and those dealing with planned change and innovations, which mention the teacher, the school as an organization, and other phenomena in the school. There are, however, some phenomena about which the present authors have been able to get very little information, for

Goals and objectives

example the school collective and group relationships in the school. Reliable information may exist, but it was not generally available when this text was written.

Most countries report objectives which imply *integrated and all-round development* of the child. A few examples will show the ways in which this may be phrased.

Spain wants to develop attitudes, habits and values of a religious and moral nature, and physical strength, and expects the child to acquire sensorial-motor skills, fitness and aesthetic appreciation. Australia mentions a balanced development of the intellectual, social, artistic and career potentials of children.

Bulgaria aims at the creation of universally developed individuals, citizens assuming active positions in life. It is also intended to stimulate the intellectual, moral, physical and cultural development of the student and to promote aesthetic education as a means for multilateral and harmonic development. In another statement, Bulgaria mentions the objective of laying the foundation for an integrated development of the child's personality; ensuring optimal rhythm for the development of intellectual, ideological and moral, physical and aesthetic qualities of the personality; and creating global ideas about environment, nature, society, home, work, etc.

Hungary states that the school shall bring about the all-round development of the personality; it should also develop physical abilities. Czechoslovakia emphasizes intellectual development and moral, aesthetic, physical and military training.

Turkey states objectives of developing abilities, motivation and attitudes. Tonga wants to achieve child development with consideration given to its intellectual, cultural, physical, emotional and spiritual environment, language and vocational development, and well-being. Malaysia aims for a balanced development of children: their intellectual, spiritual, physical and emotional development, as well as the development of their talents and the inculcation of moral, aesthetic and social values. Gabon wants an all-round education which contributes to an integrated development in the moral, civic, intellectual, artistic and vocational fields.

An objective in Colombia is to respect the development of the learner as much on the cognitive side as on the social, affective and psycho-motor aspects.

Egypt accepts the objective of a basic education that will help the children achieve integrated and well-balanced development and equip them with the bases of awareness and productive citizenship. Saudi Arabia wants an education that respects the dignity of the individual and offers him equal opportunities to develop his skills.

In some cases the objectives expressed by the different countries imply a development of the child which is in line with social and other *needs of the country*.

Spain aims for the incorporation and development of basic attitudes and desirable behaviour which enable an adaptation and integration of the child into society. The United Kingdom expresses the objective of helping children to relate to others and of encouraging proper self-confidence through greater knowledge.

The European socialist states want to enable students to become able to participate actively in social and political activities. For this reason they need an integrated ideological, labour and moral education, as well as an intellectual and physical one. Poland states that the goal of education is to prepare the pupil for contemporary life and for work in several fields of human activities so that he becomes aware of social, natural, cultural and technical realities.

Kenya states that education must develop talents and personality but must also promote social justice by instilling the right attitudes necessary for the assumption of social obligations and responsibilities. Iraq stresses the development of will-power in the human personality and the ability to participate in social solidarity efforts. The Syrian Arab Republic wants an education such that the student establishes positive rapport with his family and society, and respects spiritual and moral values and human rights.

In some cases, education is thought to develop in the learner a desirable personal behaviour, a *way of life* which is in accordance with accepted standards in the community. In connection with ideological and religious objectives, the effects of a religious upbringing were touched upon, and in connection with political objectives a few examples referring to a socialist way of life were mentioned. Reference is made to other forms of behaviour.

Bulgaria wants the school system to develop the need for a healthy way of life and to support young people in spending their leisure time in a civilized way. Education has already taken over great responsibilities for the activities of young people. Egypt stresses good conduct. The Syrian Arab Republic wants to give the student such concepts and attitudes that he/she can find his/her way in practical life or pursue studies.

The intellectual development of the child also means the development of an *ability to think* in a rational and efficient way. In a modern world, school does not mean only rote learning (if this was ever the case) but primarily critical, imaginative analysis of what shall be memorized and integration of the newly acquired knowledged into a structure of facts and theories. This has been expressed by several countries.

Spain wants to develop the capacity for imagination, observation, reflection, analysis, synthesis, etc. Pakistan wishes to create in the student an urge to observe, explore and to become aware of the surrounding world, to motivate inventiveness and creativity, and to encourage a scientific way of thinking. Guyana aims simply to help children to think and solve problems, and the United Kingdom states that children in primary school shall learn to think scientifically.

An objective in the Central African Republic is the development of a critical sense, judgement, analysis and reflection, the acquisition of rational attitudes and behaviour, and the development of memory and the ability to learn. An objective in Madagascar is to cultivate the ability to observe, a taste for research and a critical mind. The United Republic of Tanzania wants to help the pupil develop an enquiring mind and the ability to think and solve problems independently. An objective expressed by Chile is to make children think in a creative, original, profound, strict and critical way in accordance with their own possibilities. Peru wants to stimulate creativity.

In Paraguay, a stated aim is to use study techniques which favour critical thinking and reflection, and in Mexico an objective is to help pupils identify, plan and solve problems. Mexico also states the aim of enabling the students to know themselves and to develop confidence in themselves to act correctly as human beings. Senegal wants to train people who think freely, who are willing to integrate into traditional culture, and who are truthful, just, imaginative and full of initiative; to train balanced people - physically and morally strong; and to train people with a sense of responsibility, able to act, invent and create, and able to enter the twenty-first century in a spirit of scientific and technological humanism.

Guyana tries to help each child to know its own capacities and to be self-reliant. Kuwait emphasizes the importance of the individual's self-fulfilment and his need for an integrated and balanced personality development.

It goes without saying that all countries accept the objective that primary education shall prepare students for *further education*. This is expressed in various ways by different countries.

Socialist countries stress that primary school shall provide the fundamentals of polytechnical education and thus prepare for secondary education. China aims to educate children to be strong morally, intellectually and physically so as to lay a sound foundation for their secondary education. Chile and other countries point to a logical articulation between education at different levels. Mexico plays down the importance of traditional further education by stating that it is not a condition for being a complete human being; manual and intellectual work are of equal value.

Goals and objectives

Person-centred change objectives also encompass those which refer to changes in the schools themselves. Most data reported to the Conference indicate, directly or indirectly, the need for *bringing education into line with changes in society*.

A large number of countries mention recent developments in science and technology as reasons for educational changes affecting the individual. Developments and changes in society and working life are mentioned by Jordan and Sweden. In Congo the new 'School for the people' aims to bring the school culture in close contact with workers and production. The new primary school in Senegal will mean an adaptation to national reality: introduction of national language, a progressive involvement with the environment and an introduction to productive work. The new school in Cameroon will unite the former French- and English-speaking schools as well as intellectual training and work training. Bulgaria, the Central African Republic, Jordan and Seychelles point to needs for adapting the school to national realities or for integrating the school and the socio-physical environment. Czechoslovakia and Seychelles stress the importance of ensuring that learning is related to life and work and is consistent with development goals.

The changing role of the school in today's world is expressed by Belgium, which finds that primary school — for a long time mainly a teaching institution — is slowly being transformed into a school community adapted to the child's world; family/school relationships have changed and the school is more involved with social change. Egypt finds that prolonged compulsory education is necessary to prepare citizens who are in contact with social and economic reality.

The improvement of the *quality of education* is the concern of a series of responses. Mexico and Paraguay state that education does not always meet individual and social needs, and Paraguay, Tunisia and Zambia point to wastage and other aspects of inadequate systems. Pakistan wants to improve the quality of learning and cut costs. Bahamas stresses reinforcement of the quality of education since it is felt that many graduates do not master basic subjects. On the other hand, there are several countries which want to expand primary education quantitatively. Rwanda, for example, wants an education for the broad masses and find that her previous system was too demanding, and the Republic of Korea wants a proper balance between quantitative and qualitative development.

Goals and objectives of societal change

The last quadrant of Figure 3 represents societal change. Such changes can be of many kinds. They may refer to the flow of students, universalization and democratization of primary education, access of different groups to education, and so on. They can also imply that the government or the ruling party wants society to develop in a particular direction — socially, culturally, politically, or in other ways; more democratic structures, reinforcements of national culture, unified political ideas, improved economy and so on. Since education is always seen as a means for desired societal change, these types of goals and objectives occupy a prominent place in the educational debate.

Societal change goals. 'Everybody has the right to education'. This statement in the Universal Declaration of Human Rights has been accepted, at least in principle, by almost all countries. The Declaration is rather adamant and precise: on the part of every individual, primary education should be both a right and a duty; on the part of the nation, the State or the other public authority which provides such education, free education to every school-age citizen should be a duty. Different countries — which are, of course, in no way obliged to conform to this Declaration — have used varied expressions to state the principle of a right to education.

In Iraq, the Compulsory Education Law of 1978 states that primary education is compulsory for all children of the age group 6 to 12 years, and parents are obliged to enrol their children in primary school and to keep them there until they complete the primary stage. An obligation is thus placed on parents to see to it that children actually attend primary education. In Egypt and other countries, the education law states that the State is obliged to provide basic education and that parents are responsible for its implementation.

Many countries qualify this equal right to education by a statement that it is dependent on the abilities of the child or on the needs of society. Some examples, quoted from Cowen [9, pp. 16-19] may illustrate this:

The constitution of Japan enacted in 1946 provides for the basic right and duty of the people to receive... an equal education correspondent to their abilities, as provided by law. The people shall be obliged to have all boys and girls under their protection receive general education as provided for by law. Such compulsory education shall be free.

In the Jordanian statement...

1. Access to education is a fundamental right for every individual in Jordan.
2. Education is open to all school-age children irrespective of sex, race, colour or social origin.
3. Access to various levels of education or exclusion from them depends solely on individual ability and the needs of society.
4. Development of education is based on the principle of equality of opportunities.

A statement referring to the child's abilities may be interpreted differently. It may mean that special education is offered for certain pupils with difficulties, which is the case in many countries. It may also mean that different streams or different courses are made available to children at different levels of ability. In some poor countries it may also hide the fact that there is no public education for children who are not able to profit from regular teaching in school, and who are forced to drop out. As has been seen in Chapter II, dropping out of school constitutes one of the greatest problems facing education in developing countries.

The statement about 'needs of society' usually refers to the fact that a government may find it useful to limit access to education by providing only short compulsory education, by not building enough schools for everybody or not providing teachers, by charging fees, or in other ways. The political and economic realities of a nation may force it to limit the right of the individual to education. There are also a few countries which, for ideological or other reasons, do not force parents to send their children to school.

Another question which should be taken into consideration is whether educational opportunities can be made homogeneous throughout the country. In most highly developed countries only one system is allowed. This is exemplified by Sweden, where only one curriculum exists in primary school, although small local variations are found. On the other hand, countries like the United States and Australia allow differences between school districts or states which even go as far as the age of beginning school and the duration of compulsory education. Elsewhere, there is adaptation to local conditions. India reports this in the following way:

Universal Elementary Education in India caters to children of the age-group 6-14, takes education to children rather than compelling them to come to the formal system, imparts basic minimum skills of literacy, numeracy and inculcation of social and civic values; and pursues a local specific and flexible curriculum.

One of the main reasons for the introduction of compulsory education is undoubtedly the need to reduce illiteracy. An illiterate or an inadequately trained person has not only

Goals and objectives

lost his/her right to education but is also unable to participate fully in social life and does not contribute effectively to social and societal progress. One of Unesco's priorities has long been the eradication of illiteracy. The development of primary education is a main means to this end, although it has only long-term effects if it is not accompanied by adult education.

Countries where illiteracy is still common agree that a priority shall be given to eliminating it. This is stated in reports from Argentina, Chile, China, Colombia, Ethiopia, Jordan, Morocco, Mexico, Nicaragua, Pakistan, Peru, Zimbabwe and others. Reductions in illiteracy in recent years have been reported by several countries. Still, in most developing countries, a large part of all 6 to 11-year-old children do not attend school. It has been shown by a World Bank study (unpublished) that at least six years of schooling is needed to make a child permanently literate.

In all developed and some developing countries compulsory education is already complete, but cases of functional illiteracy still occur. The United States' report *A nation at risk* [50] finds that nearly 40 per cent of all 17-year-olds cannot draw inferences from written material, only one-fifth can write a persuasive essay, and only one-third can solve a mathematics problem requiring several steps. Similar provocative figures have been reported from other countries, showing that a large part of those who have attended school have not been able to profit from their schooling to an extent that makes them functionally literate and able to contribute to social development. Programmes to solve this problem have been launched. In the United States there is, for example, the National Adult Literacy Project and the Federal Employee Literacy Training Program. In Norway and other countries, use is made of the radio and TV networks, and in Sweden adult education personnel organize literacy courses for both adults and young people.

In some countries which receive immigrants from poorer countries, for example, the Federal Republic of Germany, Sweden, Switzerland, the United Kingdom and the United States, there are many new residents who are unable to cope with the languages of the host country, and a few of them cannot read and write in any language. This has called for the special provision of training for these illiterate people.

The goal of equity of educational opportunities is stressed by many countries. This may imply equal chances for boys and girls, equal opportunities in rural and urban areas and in all parts of the country, democratization of education, or discrimination in favour of disadvantaged groups.

The problem of providing equal opportunities for males and females is observed in many countries, although, as mentioned, all countries affirmed that no discrimination occurs in their responses to the Unesco questionnaire on this subject. Countries as different as Bangladesh, Denmark, India, Mongolia, Netherlands, Pakistan and the Ukrainian SSR mention that all citizens or residents shall be given education irrespective of sex, religion, race and other characteristics. However, there are countries which give priority to the education of disadvantaged people. Sweden is an example: there are specific measures to reduce, for example, sex-specific selection on choice of study and vocation.

The word 'democratization' in connection with educational opportunities is usually taken to mean an education provided without regard to pupil characteristics, such as sex, race, language, etc. It is obviously used in this sense in many country responses and in many documents referring to educational goals. In some cases, for example in countries which have recently achieved independence, the expression may refer to the fact that all citizens have recently been given, or will soon be given, equal opportunities.

Seychelles, to mention one case, plans to abolish the old parallel education system with schools of unequal status serving different population groups. Cuba uses the term differently, however, stating that education shall have a democratic character with the extension of schooling to all areas of the country, but that it shall also be democratic in the sense that everybody shall be involved in the educational process through discussions, control or attempts at solutions.

The goal of equal educational opportunities for all the population is expressed by most countries in one way or another.

The publication *A nation at risk* [50] talks about a strong commitment in the United States to the equitable treatment of all sections of the population. Finland aims to increase equity and to expand citizens' opportunities for influencing society. Greece proposes an identical training with constant quality, and the USSR states the goal of equality of all citizens to receive education. Bangladesh wants to narrow the rural/urban gap; Pakistan will pay more attention to rural and backward regions; Maldives aims to bring schooling to her many islands; and India talks about regional equity.

Some countries mention migrants. The Scandinavian countries provide the same educational opportunities to children of immigrants as to established citizens, and, in addition, language instruction in the mother tongue; in Sweden about 10 per cent of the primary school population is entitled to such bilingual-bicultural training. The Netherlands, Luxembourg and other countries give priority to immigrants and cultural minorities.

Quite a few countries stress the goals of social and societal change. In a number of cases the need for an education which supports economic and social development is mentioned. A few examples may be quoted: In Senegal an aim is to educate free men and women, capable of creating conditions for their own development at all levels, to contribute to the advancement of science and technology, and to find appropriate solutions to national development problems. In Ecuador, education is not only seen as a component in social politics but also as a component in her total development strategy. The relation between education and economic development is obvious because of the training of human resources and its contribution to scientific studies.

Chile wants to prepare the necessary human resources for the development of the country, to broaden the cultural heritage of the nation, and to assure physical, intellectual and moral development of the individual.

In Jordan, an objective is to provide the skilled technical and administrative cadres that would be able to accelerate activities in the domain of production, investment and technical development, and to improve services and project management through the best utilization of natural resources. Also mentioned is a movement towards the achievement of comprehensive development. In Saudi Arabia, a goal is to promote the economic needs of the country by ensuring that educational programmes reflect the economic interests of the community, by increasing public awareness, by modifying public attitudes towards vocational and technical training, by encouraging the participation of women in suitable professions, by associating major employers in the expansion of technical educational programmes and so on.

The needs of social development are sometimes the centre of focus. Finland wants to create conditions for improving not only the nation's standard of living but also the quality of life. The socialist countries emphasize the value of socialist education. This is exemplified by Romania, which states the following goals:

Let young people assimilate the policy and ideology of the Romanian Communist Party, scientific socialism, and dialectic and historic materialism; create and develop in young people a socialist conscience and an advanced attitude towards work, collective ownership and social obligations.

In Chile, the goal of a school which takes care of national needs is expressed in the following way: 'to ensure that school programmes are in accordance with national

Goals and objectives

reality and with the student's economic situation; to make programmes flexible so that they can vary according to regional and other needs.' In Iraq, the principle of democracy shall be established as a focal point in a well-organized life system based on freedom, equality and the rendering of equal opportunities to all. In Hungary, the school shall contribute to the openness of society and reduce educational inequalities, support personal advancement within society and reduce cultural inequalities.

Objectives of societal change. While many countries have stated goals referring to societal change brought about by means of education, few countries have actually included statements of precise objectives within this field. There are a few cases of objectives referring to the generalization of education at the primary level or to improved equity within the educational area, however, and data can also be found from educational plans.

In the countries where the whole school-age population does not attend school, there are usually plans for universalizing primary education. A number of countries even state the year when this shall be attained; for example, Chile in 1990; Ethiopia in 1999; India for children aged 6 to 11 in 1985; Madagascar in 2000; Morocco in 1990; Pakistan in 2000; China in 1990; Thailand in 1986; and Turkey for the first primary cycle in 1986. Several countries are unable to commit themselves to a precise date for complete universalization. In Pakistan, the objective is that 50 per cent of all children of school age shall attend school by 1987, and in Rwanda that 65 per cent shall attend in the same year. Several countries aim to prolong compulsory education: Iraq and the Syrian Arab Republic to nine years, Turkey to eight, and so on. The USSR and Bulgaria plan to lower the compulsory age for beginning school; the Netherlands, Belgium and many other countries will make pre-primary education compulsory; and Spain plans to extend compulsory education till the age of 16. Various steps to achieve complete enrolment in education are often mentioned (see Chapter II). They show that in many countries very high priority is given to the objective of combatting illiteracy among young people by bringing them to primary school.

The objective of educational equity is mentioned by many countries. Finland states that the generally accepted goals for education are to increase equity among her citizens and to expand their opportunities for influencing society. Denmark aims for further democratization with improved equality between the sexes and better provision for children of non-nationals. In both cases specific measures have been taken. Ethiopia wants to provide free education to the broad masses through a programme to intensify the struggle against feudalism, imperialism and bureaucratic capitalism.

In some cases limitations are reported with regard to a country's capacity to create an equitable education system. Benin mentions that the situation of the family must be considered. Ecuador states that the objectives of social justice and the consolidation of democracy cannot be isolated from plans for economic and social development: social justice implies a reduction of social differences by means of a better distribution of national wealth, but also more opportunities to obtain access to education. Certain Moslem countries have separate schools for males and females, and there may be slightly different curricula.

In most countries there are groups which can be characterized as marginal or disadvantaged in educational as well as in other respects: socio-economically deprived people; those suffering from malnutrition and similar problems; people living in remote areas; people with a different religion or culture; so-called marginal urban immigrants; national and linguistic minorities; different types of handicapped people, and so on. For economic reasons, not all countries are able to provide separate services for such groups.

However, minority groups are paid particular attention in many countries. In Australia, schools are provided with aboriginal teaching assistants who impart knowledge of aboriginal culture and form a bridge with the home. Programmes exist for expanding the number of aboriginal teachers. Teaching of the national tongue to immigrants and other minority groups is provided by most countries; special teaching of the minority language is found in some countries; and bilingual and bicultural education is reported by Argentina.

Another important group of disadvantaged children are the physically, mentally and socially handicapped. Special schools are usually created for severely handicapped children, particularly the blind and deaf-mute. Less severe cases are often found in special classes attached to ordinary schools, while slight cases are integrated into normal classes but given special attention. Great variations are found between and even within countries. In the most advanced countries, mentally handicapped children receive several years of training in special schools beyond primary school, which enables them to undertake practical work.

Poland reports a special establishment for sick children. In Seychelles all children with mental, physical or social disabilities are grouped in reorientation centres. Sheltered workshops or vocational training opportunities for the handicapped are mentioned by Cyprus, the Federal Republic of Germany, Turkey and others. Programmes for treating children experiencing difficulties with reading, writing and mathematics are found in many countries, and school psychology services are found in most industrialized ones. This shows that most countries are preoccupied with the problem of providing justice to disadvantaged groups.

Bringing education to handicapped children and so-called marginal groups, particularly in rural areas, is very expensive. Many poor countries are therefore looking for cheaper delivery systems, for example those characterized by incomplete schools, one-teacher schools, shift schools and schools based on community support. Developoed countries, on the other hand, are able to spend resources on making education equitable. This means the provision of small schools, when necessary, as well as special schools, boarding schools and school clinics.

In spite of the goals and objectives mentioned above, which aim to develop society, it can be stated that nowhere is the education system radical in the sense that it aims to change the very structure of society. The changes it is meant to support, for example by promoting universal primary education and equity, fall within the realm of the adopted general social and economic policy of the existing society.

The role of science and technology

We live in a world where science and technology almost daily bring new elements into our lives. Recent decades have witnessed the arrival of revolutionary inventions such televison, world-wide satellite communication, new printing techniques, supersonic transportation, atomic power, and microcomputers available not only in advanced countries but also in developing ones. The pace of development of science and technology seems even to be increasing.

This rapid development may pose dangers to countries which cannot keep up with it. The development activities of many of the products which are likely to shape the world of tomorrow seem to be concentrated in a few developed countries; very few countries are in a position to launch communication satellites; and the atomic reactors found all over the world are manufactured by only a few enterprises. Worse still, the new technologies seem to mean that rich countries not only become richer while the poor ones become poorer, but that the former get an economic strangle-hold on the latter. However, even in advanced economies, the new technologies are felt to pose a threat,

Goals and objectives

for example, to workers who do not feel able to handle the new work situation and by people who find it hard to get jobs.

While there is no simple solution to these problems, it is obvious that education has a role to play. In all countries, people need to learn about new technologies and new phenomena which invade their environments and change their lifestyle, and they need to know enough so that they can foresee future changes and be able to handle them. In technologically deprived areas, the role of science and technology education is even more needed than in technologically aware areas: school must acquaint students with the technologies that are not found in the environment but which influence or will influence their situation. Countries must make an effort to improve their technological status, which means training scientist and technology experts.

Under these circumstances it would be expected that the goals and objectives expressed by governments should stress science and technology in school. This is not the case. The general objectives of education rarely mention the possibility of creating education for a modern society with an important place for science and technology; the fundamental ideas are rather those of balanced development and preservation of or improvement upon what already exists.

The goals and objectives which have been classified as explicitly ideological, cultural and political imply, in fact, that the present ideologies, cultural characteristics and political ideas should be reinforced. In some cases this may mean that science and technology education receive some focus: polytechnical education, characteristic of socialist countries, means for example that science and technology, as part of society, are considered important and that the curriculum should be scientifically based. In countries based on a religious foundation, the glory of God shall be reflected in science and technology as well. However, in both cases, the need to promote science and technology is seen as subordinate to the ideological or political motive. That modern culture itself is characterized by the development of science and technology is not often mentioned, and the development of scientific and technological thinking is nowhere said to be a 'leitmotif' in the development of educational goals.

Everywhere the pedagogic goals for primary education are dominated by the development of fundamental skills, which in this context means reading, writing and arithmetic, and in some cases expressing oneself orally. Only a few countries, Sweden for example, seem to find it essential that children should also acquire skills in using simple tools found in the environment like a hammer, saw, telephone, screwdriver, electronic calculator, typewriter, etc. Even fewer countries, Denmark for example, have introduced computers into the primary school curriculum.

To a very large extent countries state that educational goals and objective shall be linked to social needs and demands. One such need is education for work, and many countries express the goal of relating the former to the latter. International conferences and international organizations, as well as individual statesmen, have repeatedly expressed this. The relationship between work and science and technology education will be discussed below. Suffice it here to quote a statement by the Conference of Ministers of Education and Those Responsible for Economic Planning of Member States in Latin America and the Caribbean, held in Mexico City [63]:

That they [the Member States] introduce policies and allocate the necessary resources with a view to:
1. increasing efforts to devise formulas and methodologies such as will ensure the compatibility of the education and production systems, with a view to increasing the pace of economic, social and political development;
2. teaching pupils from a very early age to appreciate the value of work and encouraging them to engage in work and productive activities through appropriate vocational guidance.

3. directing efforts in such a way as to ensure that marginal sectors of the population will be provided with education that will facilitate their incorporation in the labour market or their self-employment.

The reasons for attention being paid to practical work as part of education are many. In the first place, the generalization of primary education has meant that this cycle must prepare people for all types of occupations. School can no longer be isolated from the outside world, as it could when it prepared children mainly for further education. What is taught must be immediately relevant, and not only valuable for the intellectual development of the individual. It is also felt that practical work is important since it facilitates an understanding of the complex social, cultural, economic and technical conditions of the modern world. Practical work is also important as an introduction to modern science and technology. It shows how scientific discoveries are used in modern industries; it shows 'how things are done', thus paving the way for an understanding of use. Furthermore, it develops skills which cannot be taught in school. Finally, the modern world needs a positive attitude towards vocations other than intellectual ones, and this calls for factual information best acquired in a workshop.

In the case of person-centred goals and objectives of change, most countries resort to highly traditional information. These goals deal mostly with the all-round development and the multi-faceted personality of the child. In a few cases, however, a socially desirable development is called for. Of interest is the stress some countries place on the ability of children to think in a critical, imaginative, creative, scientific and otherwise productive way. Such a way of thinking paves the way for problem-solving not only in school but also in practical life and for the proper use of science and technology in different situations. This will be further discussed in the following chapters.

It is perhaps significant that the goals and objectives of education for societal change which are mentioned most often are those concerning the universalization of basic education and combating illiteracy. A look at the meaning of the term 'literacy' in different countries shows that it implies the ability to read, write and do arithmetic; some knowledge of laws, of religious ideas and of the history of the country are sometimes mentioned. No country seems to consider the concept of scientific and technological literacy; it is usually left out of the literacy debate in spite of its obvious importance to working life and to the development of the country.

Educational development, as mirrored in statements of goals and objectives, still follows very traditional lines. The old thinking in terms of disciplines seems to dominate, and there is still an unspoken conviction that school shall first provide a basis for general literacy skills, after which vocational skills may be taught. First priorities are given to literacy rather than to practical training. This may be the only feasible way of approaching educational development problems at present, but it is not the only conceivable one.

GOALS AND OBJECTIVES OF SCIENCE AND TECHNOLOGY IN PRIMARY EDUCATION

In the preceding text it was found that goals and objectives concerning the knowledge and skills of science and technology were rarely mentioned, although pre-vocational training and scientific thinking were sometimes stressed. Science and technology do appear in the primary education curricula of most countries, however, and it is therefore obvious that they are considered important.

The goals and objectives of science and technology education must obviously be in concordance with those of education in general, but the former subjects are also guided

Goals and objectives

by pressures which distinguish them from those of other disciplines. For this reason, one of the questionnaires sent to Member States as preparation for the thirty-ninth session of the International Conference on Education in 1984 requested information about policies and aims with reference to science and technology education. Various top-level international meetings, as well as meetings of international organizations, have previously expressed resolutions or recommendations on desirable standards.

Policies for and general goals of science and technology education

Regional conferences of Ministers of Education and those responsible for economic planning have on some occasions stressed the importance of adequate *science education*.

The Regional Conference for Asia in Singapore in 1971 [61] found science education was underdeveloped at that time: most pupils at all levels were almost or completely ignorant of scientific matters; teaching methods were out of date; many elements of school courses had no relationship with developmental needs, and so on. For this reason, the Conference recommended various activities to improve science teaching, for example, research, experiments and the writing of new textbooks.

The Regional Conference held in Mexico City for Latin American States [63] declared that there is a need to strengthen scientific development. It also stated that the training of people capable of introducing scientific progress and accepting the implications of their own culture is essential if they are to create, develop and adapt appropriate technologies as required by the different context of each region.

The African Regional Conference in Harare [6] found it important to develop and renew the education of science and technology at all levels for the sake of scientific and technological advancement. For this reason it was found necessary to include practical activities, which means making use of integrated science and technology teaching and the creation of suitable conditions for the elaboration of curricula, teaching materials and a pedagogy adapted to this teaching.

In the field of *technological and pre-vocational education*, the recommendations accepted by the General Conference of Unesco in 1974 provided an important input. In them it was stated that an initiation to technology and to the world of work should form an essential component of general education without which this education is incomplete. An understanding of the technological facet of modern culture, in both its positive and its negative attributes, and an appreciation of work requiring practical skills, should thereby be acquired.

It is further stated in the recommendations that the technical and vocational initiation provided by the general education of young people shall have several functions: enlarge their educational horizons by serving as an introduction to the world of work and the world of technology; orient those with suitable interests and abilities towards technical and vocational education; and promote in those who leave school attitudes and values likely to enhance their aptitudes and potentials and to facilitate their choices of occupations. Technical and vocational education should strike a proper balance between theoretical and practical work and should, among other things, be based on a problem-solving and experimental approach, introduce the learner to a broad spectrum of technological fields and to productive work situations, and develop a certain command of practical skills. It should be closely related to the local environment.

An important part of technological education is, according to these recommendations, played by practical work, and a great deal of attention has been paid to the latter in international conferences of different kinds. Several of the regional conferences brought up this aspect of training in primary school.

At the Regional Conference for the Arab States in Abu-Dhabi [60], it was said that in order to stimulate an organic interaction between education and working life, it would be necessary to evaluate the experiences made to link education to development, more particularly where education has been related to productive work, and to introduce manual work in primary and secondary education, giving it the same status as purely intellectual knowledge.

In the recommendations of the corresponding conference for Asia held in Colombo [62], it was argued that there should be a close connection between education and the world of work. Member States were recommended to integrate pre-vocational education into their curricula, to develop, as much as possible, different forms of student participation in socially useful work in the educational context, and to attach particular importance to the relationship between education and socially useful work. The pupil should not be counted as an element in the work force but as a future member of society who should learn to love work and to respect his fellow workers. This would lead to the acquisition of productive techniques and positive attitudes towards work and would, in particular, ameliorate socio-economic efficacity and give young people a sense of social and civic responsibility.

The Latin American Regional Conference in Mexico City [63] recommended that Member States initiate pupils, from an early age, to the value of practical work and productive activities, on the basis of an appropriate vocational orientation, in accordance with the physical and intellectual development of the child, and in such a way that their products become profitable for society, the school and the pupils themselves.

The European Conference of Ministers of Education, held in Sofia [7], stated that practical work forms an important part of modern education. It recommended that Member States should develop and deepen the relationship between education and the world of work and contribute by all possible means to the efforts which aim at improving the efficiency and the quality of education.

The African Regional Conference in Harare [6] also saw a very close connection between education, on the one hand, and the economic, social and cultural life of the community on the other, that is, between theory and practice. An intimate relationship between the acquisition of knowledge, skills and attitudes and productive work was found necessary if education was going to play a significant role in the transformation of the environment and the community it serves. Productive work contributing to the self-financing of educational institutions was mentioned. A series of recommendations implied integration of productive work into educational activities while taking national realities and characteristics into account. Stress was also placed on the need for a revision of curricula at all levels and for the preparation of educational materials which increase the educational value of productive activities.

The thirty-eighth session of the International Conference on Education, held in 1981 [29], had as its main theme the interaction between education and productive work. A pre-conference questionnaire was sent to Member States, and the replies implied that governments gave increasing attention to the interactions between education and productive work and between educational activities and the world of work. Some governments reported that one reason for this was to avoid the traditional isolation of the school, the result of an undue concentration on intellectual development by many educators and a lack of understanding of the complex social, economic and technical realities of the world of work. Other answers referred to production. Evolving scientific applications and technologies require changes in the knowledge and skills taught in school; no sharp distinction should be made between theory and practice in training and education; an interdisciplinary approach becomes natural.

No country or international organization negates the importance of technology in the primary and secondary school. Based on the answers to the questionnaires sent out in connection with the thirty-ninth session of the International Conference on Education, it is also quite clear that the school system should be responsible for an introduction to technology. The reason is the increasingly large demands being placed on citizens by the present world development. An interesting development has been found in Sweden but is certainly relevant to other countries: many students have lost interest in purely theoretical courses in school, which are found to be of little practical use to them. The manipulation of technical equipment, however, ought to raise questions in the students' minds about the theoretical background to the phenomenon in question [14].

In this connection it should be mentioned that not all countries agreed on the creation of a close link between education and productive work or to the inclusion of pre-vocational training in the primary school. The thirty-ninth session of the International Conference on Education declined to include recommendations of this nature. International teachers' organizations also expressed concern at this conference session about the inclusion of practical work in the curricula for other than purely pedagogical reasons [30].

The questionnaire mentioned above, sent out in 1984, invited Member States to indicate whether they had accepted any *policy in respect of science and technology education*. A few reported that they either did not teach these subjects at the primary level or that no overt policy existed; others referred simply to the fact that the disciplines belonged within a general undifferentiated subject, as will be discussed in the next chapter. Still others did not answer the question.

Science and technology form an essential part of polytechnical education, and a close relationship is created between school and practical productive work; this general policy is referred to by several socialist countries. The USSR mentions that the strengthening of labour education and technical knowledge and skills is established policy for the period up to 1990. The Byelorussian SSR's school has a unified labour and polytechnical curriculum. Hungary refers to the work by the Hungarian Academy of Science to outline the content of general education; it describes an ideal with cultural material embedded in science and technology education and other subjects. In Bulgaria, there is a definite policy for science and technology education at the primary level. It provides the foundations and permits the child to assimilate science and technology and to acquire certain skills and capacities. In Czechoslovakia, the educational plan tries to strike an optimum balance between all the different aims of education, among them labour education and personality formation. Other countries, such as Algeria, Ethiopia and Cuba, refer to their polytechnical education. The last-mentioned country states that education shall create a scientific concept of the world and encourage independent work habits at the same time as developing a harmonious, all-round personality.

The Mexican national development plan 1983-88 gives priority to the following areas at the primary level: revision of curricula for the integrated training of pupils in science and technology; stimulation of technological culture by improving technical education from the primary level; and improved teacher training. Nepal and Zambia, among other countries, give priority to science and technology in primary school. The Netherlands is pursuing a definite policy with regard to the teaching of these subjects and refers to projects under way for the creation of work plans for nature studies in primary school. A policy, expressed in the United States publication *A nation at risk* [50], but certainly shared by many countries, is presented in the following way:

Knowledge of the humanities... must be harnessed to science and technology if the latter are to remain creative and humane, just as humanities need to be informed by science and technology if they are to remain relevant to the human condition.

This analysis of declarations made at important international conferences and by individual countries indicates clearly that more and more countries accept the important relationship science-technology-productive-work in primary school and that they also see education as an integral part of a greater whole — an institution in the service of the larger society. The old concept of education as an isolated phenomenon which stated its own goals and was characterized by compartmented disciplines left little room for technology and none for practical work. This change in philosophy has been most pronounced in countries which have adopted the system of polytechnical education. It has also been rather important in Western countries where the idea of an education related to the surrounding world was propounded by philosophers like Rousseau and Dewey, as well as by more modern educators such as Piaget, Brunet and Vygotsky. It is also becoming accepted in other areas of the world. The idea of an education adapted to local and national conditions, as opposed to the idea of a school based on a traditional 'Western' model, is also slowly making headway according to the available policy documents.

Objectives of science and technology education

The model of educational goals and objectives given at the beginning of this chapter will also be used to describe the objectives of science and technology education in school. Such objectives can be gleaned from many different sources, of which the main one is the questionnaire referred to above.

No differentiation will be made between goals and objectives in the text below. In the case of aims for the teaching of a particular subject in school, authorities are likely to express themselves in fairly concrete terms, that is, by stating objectives; it is not common to find vague goals statements in the material available. Some of the objectives are primary ones, that is they express the aims of creating certain types of behaviour on the part of the individual, the class, the school system or society. In other cases, they are secondary ones, in the form of demands for educational products like knowledge, skills, attitudes, values, etc.

Explicit ideological, cultural and political objectives. Undoubtedly, many people have found it difficult to perceive a relationship between the ideological, cultural and political needs of society on one hand and the teaching of science and technology on the other. Ideologies have often been conceived as systems of ideas and beliefs; culture has primarily been connected with aspects of civilization like literature, fine arts, religion and humanistic schooling; and political objectives of the kind discussed in this section — patriotism, the propagation of political ideas, and so on — have had little to do with practical and theoretical training in science and technology.

That these are very constraining conceptions has, however, become more and more obvious to educators and politicians all over the world. As has been discussed, polytechnical education finds it essential to relate school to society and to apply what is learned in practical situations. Moslem and other countries have expressed, as a general goal of primary education, that they want to use the development of science and technology to show the glory of the Creator. It is realized that scientific thought and technological innovations form integral parts of culture. As is well known, scientific and technological developments also enter into political programmes.

Some countries answering the questionnaire have stated goals implying the deepening of their culture through scientific and technological development. Sri Lanka wants to cultivate among her people an appreciation of the value of science and scientific methods as an indispensable part of modern society. Nicaragua wishes to deepen the roots of her own culture and profit from the scientific and cultural development of mankind. The United Arab Emirates aim to create a deeper belief in and respect for the

Creator by studying creation in detail. Hungary and Kuwait mention science as part of modern culture.

Slightly different objectives are expressed in the answers from Turkey and Gabon. The former country states that technological education is important because it allows the Turkish nation to gain access to industrial culture, and the latter wants to initiate the pupils into traditional technologies in favour of an endogenous development, to safeguard cultural identity and avoid depersonalization by tying education to the realities of the environment.

Educational objectives. In its *Sector policy paper*, the World Bank [11] stresses the importance of education to man. It is found that education meets a basic human need by providing a broad basis of knowledge, attitudes, values and skills on which to build in later life. It is also said to influence and to be influenced by access to other basic needs. Among those mentioned are nutrition, safe drinking water, health, shelter and prevention of disease. Finally, education is found essential for preparing and training manpower to manage capital, technology, services and administration in all sectors of the economy. All these points indicate the need for a science and technology training which provides knowledge, skills, attitudes and values.

Eriksson [14] shows that science and technology education is essential since it transmits knowledge and skills which pertain to modern culture and form part of our cultural heritage. Economic and social changes in society during the nineteenth and twentieth centuries have essentially been based on contemporary scientific and technological developments. Many people distrust this development; others invest it with utopian powers. In order to allow people to enjoy its fruits and to judge its positive and negative aspects, fundamental knowledge is absolutely essential.

A large number of countries have stressed that science and technology education shall provide *knowledge and understanding* of science and technology as a whole or of specific areas of it; in many cases reference is made to the environment. A few examples of statements may show this:

The German Democratic Republic provides a general education, common to all, including knowledge from all essential spheres of culture. It forms the basis for any creative activity, for any post-school education, and for the growth and changing demands made by working life. Polytechnical education is part of general education. Scientific and technological education at the primary level includes the familiarization of children with their locality, socially useful work which is geared to the children's age and an introduction to fundamental principles. Poland initiates the child to the structure and evolution of the world at the micro and macro levels, thereby assimilating the principal laws and theories which serve as a basis for the formulation of scientific views of natural phenomena. In Hungary an aim of science and technology education is to present general principles, like causality, inertia, macroscopic qualities determined through microscopic qualities, etc. Science education is used to give a window on the world and to form a global scientific world view. Technology education creates an interest in technical work and present technological knowledge as an integral part of general knowledge.

Mexico wants the pupil to understand how equipment functions and is maintained; to be able to understand and apply simple techniques and technological processes, both modern and traditional; and to notice how technology influences life.

The fundamental objective in Iraq is to enable the pupil to develop scientific skills and abilities, and an understanding of using technological equipment in life, whether at school or at home, but also to protect the environment. As far as polytechnical education is concerned, Algeria stresses the acquisition of the fundamentals of science and technology in order to analyse and comprehend the living and inert environment, and mention is also made of knowledge about production processes, and an education for and by work. The Syrian Arab Republic wants certain ideas to be taught so as to enable the pupil to understand the environment and science around him and in order to continue his studies at the secondary level.

Kuwait aims to provide knowledge of science and scientific methods which allows an understanding of the environment, including the fundamentals of scientific research and the role played by science in society and modern civilization. It is also intended to give the pupil complete and consistent scientific explanations allowing an understanding of new discoveries and to know the general basis for the technological realization of scientific discoveries. The United Arab Emirates mentions knowledge of method, knowledge of the environment and its protection, and knowledge of scientific discoveries affecting daily life.

Rwanda wishes to train individuals with both theoretical and practical knowledge which allows them to integrate successfully into their environment, particularly the rural one, and Nigeria wants science education to develop functional knowledge of concepts and principles, and explain simple natural phenomena. Cuba stresses that the knowledge acquired should reinforce a scientific concept of the world and that it should develop independent work habits.

Peru mentions adequate knowledge of the principal phenomena in nature with special reference to local and national reality, and Argentina states that on the basis of its natural curiosity, the child shall be initiated to scientific activities which lead to an understanding of the natural world.

Cultural and general education, with the stress on training citizens to acquire technical knowledge and skills, is an aim in Turkey. The schools in Colombia should teach basic concepts about the structure of the universe and of interactions between and within systems; mankind, as a rational and living being, should conform to interacting systems in order to maintain a biological and social balance.

China stresses the task of guiding children into some preliminary undertaking related to nature, its exploration, renewal and conservation by mankind, thus helping them to acquire useful knowledge about the natural sciences. Knowledge and working skills shall be acquired by means of observation and experimentation, and children's interest in and ability to learn and use science shall be developed.

An objective in Mauritius is that children of primary school age should be helped to quantify their ideas and to develop and utilize spatial concepts, nutrition practices, etc., in their daily life. Kenya stresses practical skills and the development of positive attitudes towards work. Another area of emphasis is the introduction of science and technology teaching so that pupils may begin to understand the physical environment and the application of science and technology in the solution of problems and the accomplishment of tasks.

Seychelles wants to inculcate knowledge and an understanding of scientific and natural phenomena, as well as an ability to communicate with others about these subjects. The United Republic of Tanzania would like to develop children's curiosity and ability in arts and crafts, and provide them with the basic principles of science and technology in accordance with their environment. Australia aims to develop knowledge and an understanding of chosen content within the biological and physical sciences. Cyprus wants children to understand simple concepts, principles and laws, and to develop positive attitudes for the appreciation of the environment.

In a few cases, countries refer to specific subjects or areas in their answers.

Tunisia wants students to acquire fundamental concepts of mathematics, physics, chemistry, biology and technology. Luxembourg mentions a new manual of geography and materials for sex education. A few countries refer to studies of the immediate environment or of nature in general.

Goals and objectives

These objectives of knowledge and basic understanding have also been stressed by other studies. The Meeting of Experts on the Incorporation of Science and Technology in the Primary School Curriculum [47] stresses that science can equip future citizens for an increasingly scientific and technological world, and that it offers an opportunity to explore their environment logically and systematically. The study *Technology education as part of general education,* conducted by Unesco [84], stresses that technology education imparts essential practical knowledge to students and develops proficiency in the use of tools and equipment. It quotes a statement from the United Kingdom, which may be cited here, too:

Major aims of technology education: to develop in all students a general understanding and appreciation of and a sympathy for scientific and technological development; to enable all students to become involved in at least some parts of the technological design process.

Life problems, for example those which have to do with energy, survival, understanding of life, and living and growing up, are brought up by a few countries.

Paraguay aims to describe the biological structures and functions that intervene in the co-ordination and relation of organisms with the environment. The child shall be made to understand the basic biological processes of nutrition, growth, development and reproduction. Thailand wants the child to acquire a basic understanding of and appropriate behaviour concerning mental and physical health, at both the individual and community levels, and to possess basic knowledge and understanding for survival. Malaysia wants children to acquire some basic knowledge and understanding of mankind and the environment, and of the interaction between the elements of nature.

The acquisition of *manual skills,* positive attitudes towards manual work, and skills useful in practical life is referred to in a few answers. As mentioned above, Mexico listed the functioning and maintenance of equipment, and the United Arab Emirates and other countries report work skills as a goal.

This type of goal was alluded to in the 1983 Unesco study of technological education as part of general education [84]. It was expressed as the development of a sense of dignity of labour, the development of systematic, tidy and safe work habits, the acquisition of knowledge which can be used in the world of work, the development of proficiency in the use of tools and equipment, and so on. Denmark stated:

Many children will, however, by the satisfaction they feel in working in the school's workshop, get interested to choose an education and a job which involves manual skills combined with a certain creativity.

The acquisition of *positive attitudes* towards science and technology is mentioned by several countries, for example, Australia, China, Colombia, Cyprus, German Democratic Republic, Guyana, Hungary, Nepal, Paraguay and the United Arab Emirates. This aim has been described in different ways, for example: Guyana — to develop desirable values and attitudes towards science; Nepal — to create an interest in studying nature and the environment, and to appreciate them; Australia — to develop positive attitudes towards the responsible use of science and technology; German Democratic Republic — to initiate scientific and technological interest and the desire to look into and understand the phenomena and processes in nature and technology; and the United Arab Emirates — to create positive attitudes towards science and technology, and respect for manual work.

The importance of creating positive attitudes towards these subjects has been stressed in other reports. The IEA study [27] describes relatively positive attitudes towards science and scientific development. Among the objectives of technology education reported in the 1983 Unesco study of technology in general education, some had to do with interest in the subjects or their application in practical work.

Knowledge and *understanding of scientific and technological processes* is an essential aim of science and technology education, according to a large number of countries. The latter point to the importance of teaching students to observe, experiment, analyse and interpret data and draw correct conclusions from the material.

Australia wants children to develop the attitude of scientific inquiry and scientific skills. They should also gain insights into technology, its relationship with science, and the way in which science and technology have fashioned our environments and lives. The United Kingdom stresses the importance of introducing children to the skills and processes of science, including observation, experiment and prediction. Primary school should therefore seek to develop such processes of scientific thinking as observing, pattern-seeking, explaining, experimenting, communicating and applying. The need for actually carrying out scientific procedures is stressed — not just reading or hearing about it. In Belgium, particular attention is given to the formulation of observations, to relations between experiences, and to scientific attitudes like exactness and objectivity. France states that the subject *Activités d'éveil* should aim to initiate observation, experimentation, measurement, systematization and documentation. Cyprus realizes the need for teaching children a scientific approach to solving problems. Hungary expresses the aim of applying scientific methods to discover the world by providing objective experiences, depicting the essential and creating a model; and by prognostication, control, correction, further development of the model by testing and application of the model.

Nicaragua aims to promote the student's scientific, critical and reflective attitudes while developing skills, knowledge and dexterity which allow them to incorporate themselves in a creative way in the world of work. Brazil wants the development of logical thinking and scientific methods without neglecting the techniques that result from their application. Paraguay mentions the application of basic scientific processes in real situations. In Guyana, an objective is to develop the ability of the child to think clearly, to discover significant scientific facts and concepts, and thus to arrive at generalizations. Nepal wants the students to develop the habit of generalizing on the basis of scientific experiment and thinking in a scientific way, while Seychelles mentions their ability and willingness to use a logical and rational, 'scientific' approach to everyday life based on objective observation.

Japan stresses the goal of developing children's abilities to carry out scientific observations and to deal with natural phenomena in daily life, and Mauritius states that pupils should be able to observe and formulate precise questions relating to the natural and social environment, and to interpret and communicate their findings. Malaysia expresses the goal of developing skills in thinking, reasoning, inquiry, evaluation and decision-making concerning mankind and the environment. Gabon mentions work methods like documentation, measurement, integration, application, etc., and the use of tools, carrying out repairs and so on. Nigeria states the objective of acquiring basic scientific skills: observing, manipulating, classifying, communicating, inferring, hypothesizing, interpreting data and formulating models. Uganda wants to help the child develop skills of observation experimentation and evaluation.

The answer from Iraq is expressed in the following way:

Develop the pupil's way of thinking towards a more scientific attitude from an earlier stage in order to be able to solve their social, parental, and everyday life problems by putting possible solutions to face such difficulties... a scientific point of view which aims at preparing the new generation to be more responsible, have a better way of thinking scientifically to be able to face difficulties in the future.

Jordan emphasizes the importance of adopting a scientific method of thought and of developing creative abilities. Kuwait strives to create habits of thinking in all judge-

Goals and objectives 85

ment situations, and wants to help the pupil acquire a scientific method of thinking showing a belief in reason, a broad view, freedom from superstition, an ability to avoid making decisions on an unsatisfactory basis — a belief that scientific facts may bring changes and an ability to avoid generalizations. Tunisia wants children to acquire a scientific spirit progressively: reflection, observation, experiment, analysis and synthesis.

The aim of making pupils understand that science evolves is expressed by some countries. Colombia wants them to realize that scientific knowledge is not definitive but is in constant evolution in accordance with the culture of each era. They should know that the learning and application of scientific methods allows all people to participate in the improvement and development of knowledge. Mexico wants them to understand that science encompasses both the existing knowledge of nature and the search for new knowledge. It is necessary to keep the natural environment under constant observation using basic scientific processes.

This necessity for the scientific training of children was also observed by respondents to the 1983 questionnaire on technology in general education [84]. They stated the need for encouraging students to explore and experiment and to develop the theory, practice and science underlying work principles, etc. The Meeting of Experts on the Incorporation of Science and Technology in the Primary School Curriculum [47] mentioned several objectives of this nature:

- to develop scientific attitudes, interests and processes that will help improve the quality of life;
- to develop an inquiring mind and a scientific approach to many situations;
- to develop an ability to interpret information critically, and to evaluate different alternatives when making decisions;
- to develop the ability to acquire knowledge, to think logically about it, and to conduct simple research; and
- to develop skill at communicating effectively and discussing rationally with others.

A few countries have also expressed the objective of making children develop a *scientific attitude*: curiosity, criticism, search for truth, and so on. This attitude is thought to make them willing to devote themselves to the pursuit of scientific and technological activities. This is expressed somewhat differently by different countries:

Australia wants to develop the attitude of scientific inquiry and skills; Belgium states that primary school should develop critical and objective thinking, not presenting definitive knowledge but awakening a sense for observation, curiosity, a search for scientific relations, and questioning of interpretations and tentative solutions; and France aims to develop a scientific attitude of curiosity, as well as creativity, a critical sense and the need for objectivity. Nicaragua wishes to promote scientific, critical and reflective attitudes in the students, and tries to stimulate in teachers and students the capacity for critical and self-critical, participative and creative analysis which turns the school into a liberating force.

Morocco mentions the need to develop initiative and creativity in the child. The Republic of Korea aims to cultivate intellectual curiosity, creativity and research attitudes, an ability to solve problems rationally and originally, and an ability to apply fundamental scientific knowledge and skills. Gabon wants to draw out curiosity, critical thinking, creativity combined with responsibility, co-operation and team work. The United Republic of Tanzania wishes to develop a sense of discovery and a creative attitude in children. Mauritius also aims at the development of an inquiring mind and a scientific attitude. Rwanda expresses the aim of developing the child's intellect by initiating sound reasoning abilities, a critical spirit, precision and creativity. Malaysia, finally, strives to develop understanding, consciousness and sensitivity towards the causes and consequences of changes in society.

The final type of curriculum objectives which have been expressed by different countries are those of allowing the child to *learn by observing phenomena in daily life* and deal actively with them: to become acquainted with the environment; to collect data from it; to study nature; to acquire information about phenomena like energy, forces in the social and natural milieu, and health; and to investigate Man and his society.

Egypt wants to acquaint the pupil with the environment; first with school, family, food and food sources, the local district, etc.; then with what is going on in the surrounding world and with the relations between the human being and the environment. Tunisia states the objective of making the pupil learn about mankind's place in nature, his ability to transform nature, and the influence of technology over the environment.

The United States refers to a report by the National Science Board which proposes: 'Science and technology education daily in every precollege year with emphasis on phenomena in the natural environment, collecting and processing data, and a balanced physical and biological science program.' The Netherlands mentions that a project exists with the aim of preparing plans for the teaching of nature studies in the lower forms of primary school. The United Kingdom states that science education should include the study of living things and their interaction and deal with the environment, with materials and their characteristics, with energy and its interaction with materials, and with forces and their effects. The teaching of science should lay the foundation for a progressive deepening and understanding of scientific concepts and facts. The content of science should, whenever possible, be related to the children's experiences; it should provide them with knowledge of scientific ideas which would help them understand their own physical and biological environment and understand themselves.

San Marino states that science education aims to present the pupils from their early childhood with scientific problems and techniques related to real life. France finds that the child should slowly be given a possibility, starting from problems in the immediate environment, to organize a set of skills and knowledge which brings about an understanding of the world in which he/she lives. Sweden states that everyday knowledge must play an important part in school. This includes knowledge concerning the home, family life, interpersonal relationships and technology; and more traditional subject matter may have to be sacrificed in order for the school to relate sufficiently to the reality surrounding the pupil.

Japan states the objective of developing children's ability to conduct scientific observations and deal with natural phenomena taking place in daily life. Turkey intends that the pupils shall first know themselves and their environment. Seychelles wants children to develop the ability and willingness to look at and observe their surroundings objectively. Nigeria wants children to observe and explore their environment.

Understanding phenomena in the surrounding milieu means, in fact, that two types of goal are kept in focus. The environment is an area of knowledge. The publication issued after the Meeting of Experts on the Introduction of Science and Technology in the Primary School Curriculum [47] stated this in the following way:

The items of the curriculum content will be selected in some countries by taking into consideration that primary education is often an end of formal education for many children. Thus it may include simplified knowledge about themselves, nutrition, environment and its phenomena, principles of basic science knowledge (biological and physical sciences), physiology of some human systems, the principles of hygiene, and the conservation and rational uses of the environment.

The study of the environment is also a means to inculcate theoretical aspects of science and skills useful in other areas. It means starting with familiar phenomena and proceeding to less well-known ones and to generalizations. This conveys particular importance to this area.

Objectives of person-centred change

There are few modern educators who would not insist that an essential aim of education is to bring about changes: in the individual learner and often in the teacher, and in groups and group relationships. This was expressed by most countries in respect of primary education at large, and it was found that the all-round personality development of the child is considered a major goal while very little could be concluded about other types of development. Most countries also found that science and technology should play a role in this development.

Some countries express the belief that this education can *change the child's personality*, that is to say, influence its socio-emotional and intellectual growth, the development of abilities, and help it to become self-confident.

Paraguay wants the child to adopt positive attitudes towards work as a condition for self-realization and a source of income. Cuba stresses that science and technology education shall inculcate cognitive capacities in accordance with the purposes of education in the country: the creation of a multilateral personality which is harmoniously developed by means of an integrated education. Jordan wants to develop the intellect of the child to improve its understanding and memory of information.

Mauritius wants an appreciation of relationships: children should learn how to live collectively, to show respect for various cultures and religions, to show respect for rules and laws, and so on. In Malaysia, an aim is to develop and inculcate positive attitudes in the application of knowledge and skills towards solving problems and issues concerning the individual, society and the environment, and to develop and inculcate values and positive attitudes towards living together in harmony in the context of Malaysia's plural society. The Republic of Korea teaches technology to stress the value of the pupil's own existence and the desire to become self-reliant with all the necessary abilities. It is also intended to develop self-reliance through co-operative involvement in work.

Nigeria strives to develop self-confidence and self-reliance through problem-solving activities in science; and to develop a functional awareness of and sensitivity to the orderliness and beauty of nature. Uganda wants to help the child to develop self-reliance, resourcefulness, problem-solving abilities, and experience of scientific methods. Ethiopia, finally, aims to let the child acquire relevant and fundamental science and technology education for self-reliance and self-sufficiency.

This goal was also shared by the Meeting of Experts mentioned above [47], which stated that science can help children to think logically, speed up the child's intellectual development and, through its applications, can improve the quality of people's lives.

A few countries mentioned another aspect of the pupil's personality development, namely learning to *appreciate nature* and being prepared to live in close contact with it.

Poland wants pupils to be familiar with the natural values of their native country and appreciate the beauty of its landscape. Japan mentions curriculum content which takes into account children's mental and physical development and which cultivates their love of nature. Guyana wants to draw out desirable values and attitudes towards self and others, the environment, the country and science; and Nigeria intends to develop a functional awareness of and sensitivity to the orderliness and beauty of nature.

The development of suitable *attitudes towards life* is also mentioned in a few answers: personal welfare, responsibility, the use of knowledge in everyday life, attitudes towards reality and so on.

Hungary aims to arouse a love of work, a feeling of responsibility, discipline and proper attitudes to work. The Ukrainian SSR wants to develop the pupil's personality, condition his attitude towards reality and strengthen his character. Colombia mentions in connection with technology education that it shall promote the development of initiative and creativity by introducing new methods, etc. It shall also shape positive attitudes and values in respect of personal, social and environmental safety and hygiene, and in respect of the best use of services and welfare.

Cameroon wants children to acquire an ability to assimilate innovations and teaches them to adapt to the life that awaits them. In the Central African Republic, an objective is to integrate the child effectively into the environment and into modern life. Nigeria and Seychelles talk about ways of applying the knowledge and skills acquired through science in everyday life.

An *interest in the environment*, which expresses itself as a will to observe and interact with it, and a desire to adapt to nature and the social milieu, is mentioned by a few countries.

Turkey finds that children shall first learn to know themselves and their environment. Science education can play a large role in the development of the capacities of the pupil, and technology education is important because it aids him in adapting to his environment. Thailand wants the child to adapt to a changing environment and to use scientific and technological skills in daily life. It also wants the child to appreciate the relationship between the individual and the natural, technological and social environment. Seychelles aims to create an awareness of the influence that science and technology have both on our everyday life and on the lives of other people, plants and animals.

Colombia states the intention of applying scientific techniques and knowledge for the solution of problems related to the individual's health and the preservation of the natural environment. It is also intended that children should value scientific knowledge and technical innovations as expressions of the human capacity to interpret, transform and make use of nature. Jordan endeavours to enable children to adjust to the environment in which they live. Cyprus wants to develop positive attitudes towards and appreciation of the environment and to get children involved in its conservation.

The Meeting of Experts [47] stressed this goal. Children should, according to them, develop an appreciation of the patterns and relationships that exist in nature and the methods by which they can be studied, and they should develop confidence in their abilities to face difficulties. Education shall be relevant to the social and environmental context of children.

The last aspect of person-centred change objectives mentioned by various countries is to prepare the child for *leaving primary school*.

Further education at the secondary level is briefly mentioned by, for example, Algeria, Iraq, Kuwait, Rwanda, the Syrian Arab Republic and the USSR, and the importance of enabling the child to choose a vocation or a career by Hungary, Poland and the Ukrainian SSR. Neither of these objectives has attracted much interest. The former is self-evident and does not call for comment, and the latter is probably not felt to be essential except in countries with polytechnical education, since primary school is not usually considered as preparation for an occupation or a vocation elsewhere.

Societal change objectives

It is generally thought that education can bring changes — usually improvement — to society. The building up of existing education systems, which has been going on for a long time and which has accelerated since the Second World War, was founded on the assumption that education can bring about development, particularly in the economic sphere. It was seen as an investment. Many people still see this relationship, and quite a

Goals and objectives 89

few countries have mentioned objectives of societal development underlying scientific and technological education.

Training for *creative and productive work* was found to be an objective by some countries, particularly those employing polytechnical education.

Uganda mentions the objective of teaching the child to be productive by acquiring basic work skills. Nicaragua wants to guide the student towards productive work by promoting new attitudes and by making use of the information and the scientific and technological knowledge which has been provided. Algeria mentions preparation for and by work. Poland states that technical work is the guiding principle behind polytechnical education.

Preparation for active life was referred to by the Meeting of Experts [47] which stated:

Relevance [to real life] was essential, since science is difficult for many primary teachers, and their motivation for making the necessary effort must come from the conviction that science is as much a part of preparation for life as is numeracy and literacy. One aspect of relevance was, therefore, considering the future of the child in the society in which he or she lives. The child must be equipped with skills and attitudes for coming to terms with the influence of technology in every day life and with the 'knowledge explosion'.

The Unesco study *Technology education as part of general education* [84] mentioned this aim. The pupil shall be prepared for making useful contributions to life; he shall learn to appreciate the roles of agriculture and industry; he should be guided towards the world of work, and so on.

The influence of science and technology education on the *economy of the country* by supporting production, teaching people proper consumption habits, providing skilled workers, making people aware of the need for their work, and so on, has been stressed by a number of countries.

Nicaragua states the objective of providing humanistic, scientific, technological and other training which will transform reality in a creative way and will bring forth the human resources, both technical and political, necessary for the progress and consolidation of the revolution. Jamaica aims to raise national awareness of scientific literacy and technology, and to create citizens capable of applying appropriate scientific skills to the solution of problems. On the subject of technological education, Colombia hopes that the functional linking of formal education with active work will lead to personal and social development. Paraguay wants the citizen to apply appropriate technologies for production in agriculture and forestry, and to take part in the work of conserving the soil. It is also intended to inculcate beneficial attitudes to the selection, preparation and consumption of food.

Morocco wishes to teach the pupil the importance of productive work through science and technology. An aim in Sri Lanka is to ensure that education and research institutions produce scientists and technologists of the highest calibre, and another is to make available as widely as possible within the country the fruits of scientific and technological activities. Turkey wants to teach the primary school child the abilities called for by the national awakening, since the need for qualified manpower is felt. Technical education could aid the trainees to develop their capacities to produce, live a healthy life, learn to do research, co-operate and communicate with others.

Viet Nam mentions the task of carrying out the three revolutions: production; science and technology; and ideology and culture. The second is said to be the key one. Poland states that work training in school contributes to better knowledge of problems in production, to a better understanding of the importance of work in personal and social development, and to vocational guidance. The Ukrainian SSR wants to satisfy the need for specialists and qualified workers. Switzerland states educational objectives in the following way:

In a country like Switzerland, which has acquired its wealth thanks to its economic development, they are very conscious of the importance of science and technology education in compulsory school but it is primarily during the last three years of this education that science and technology training is systematically organized (ages 12-15).

In the United Kingdom, the government sees the development of science as an important feature in the economic and social life of a country where technological and scientific applications have great significance.

The *International yearbook of education*, Volume XXXV [22], brings up this need. It is stated that education shall ensure economic growth and prepare well-qualified men and women for the labour force.

In many parts of the world, there has been an intense discussion of the need to *preserve natural resources*, particularly in energy. This is reflected in a few statements.

Mexico states that the child shall participate in an adequate way in the constructive use of scientific knowledge for the improvement and conservation of the natural milieu, and Paraguay wants its citizens to co-operate in activities tending to conserve the country's natural resources. Hungary wants education for environmental protection and good management of material, energy and time. Australia mentions that a main objective of science and technology is to equip primary school pupils for constructive interaction with their environment (both natural and man-made), and Mauritius states that the pupil should learn to avoid wastage of natural resources and pollution of the natural environment.

The movement for the conservation of resources is discussed by many authors. Eriksson [14] discussed the need for training of this kind within the framework of science and technology. One of the goals mentioned in the Unesco study *Technology education as part of general education* [84] was to teach students to avoid waste and to conserve, utilize and develop natural resources.

Other change objectives have been mentioned by some countries: co-operative training, team work, equal opportunities for all, democracy of training, a better quality of life for all, respect for public property, and so on. In general, however, the objectives mentioned concern aspects of the economic development of the country. This fact will be further discussed below.

The different roles of science and technology education

The principal question to be raised here is whether countries have accepted the potential of science and technology education to improve the primary school and to influence the development of the country as a whole, or whether these subjects are given the same priority as conventional goals and objectives.

The study of the *ideological, cultural and political objectives* of primary education as a whole gave a rich variety of such aims; primary education is apparently seen as a means to strengthen the ideological and religious foundations, the cultural heritage, and the political and patriotic basis of a country.

Few countries have stressed such goals and objectives in respect of science and technology education although some countries have mentioned objectives like strengthening the national culture, profiting from scientific and cultural development, and using traditional techniques. Reference has been made to the fact that science and technology form part of modern culture, and countries with polytechnical education are adamant in their search for a natural relationship between what is taught in school and what goes on in the environment. A true realization of the cultural and religious values of science

Goals and objectives

is expressed by Moslem countries when they see it as an expression of the glory of the Creator

The cultural and political features of a nation or of mankind are expressed not only through history, literature, religious development and so on, but also through scientific involvement and development and through the application of technologies in industry and daily life. No country has made this clear. The fact that scientific discoveries and technological innovations have changed the very structures of the social and humanistic sciences, exemplified by the use of statistical methods and data-treatment techniques even in purely humanistic research, has evidently not been considered. To this can also be added the facts that the boundaries between the different disciplines are breaking down and that integrated or interdisciplinary approaches are becoming more and more important.

A rich variety of *educational goals and objectives* of primary education were mentioned by Member States. The most essential ones were those of literacy and numeracy, while science and technology were rarely given priority if mentioned at all. Such goals and objectives can be either primary ones, expressing expected behaviour, or secondary ones, expressing products of education like knowledge, skills and attitudes [91,p. 116 *et seq.*]. Both types of objectives were referred to in respect of science and technology education.

Primary curriculum objectives of science and technology education are common. Most countries have referred to the acquisition of scientific skills in general or to some aspect of them: observation, experimentation, prediction, pattern-seeking, analysis and synthesis of facts, communication, application, classification, logical thinking, measurement, interpretation, formulation of questions and hypotheses, search for essentials, testing, model building, control, generalization and so on. Quite a few countries have also pointed to the essential objective of developing a scientific attitude: broad view, freedom from superstition, belief in facts, curiosity, criticism, search for truth, objectivity, creativity, reflection, self-criticism, originality of thinking, team spirit and so forth. Exploration as a means towards an understanding of science and technology is mentioned. This includes nature study, learning about and from the environment, and a will to observe and study the latter.

The skilful application of science and technology in daily life, for example for the purposes of nutrition, health and shelter, is stressed. Practical skills in the handling, storage, and maintenance of tools and equipment, and skills in production processes are found essential by a few countries. The ability to adapt to and integrate into the environment is also mentioned. Finally, a few countries mention the goal of continuing education.

The classical role of science and technology teaching has been to let students acquire knowledge of scientific facts. This is another role to be played by modern science and technology. Different countries have expressed the objective of scientific and technological knowledge, understanding and skills differently: as knowledge of fundamental principles, laws and theories; as general knowledge of science and technology; as comprehension of science and technology; as an ability to understand new discoveries; as the acquisition of scientific and technological skills; or as knowledge and understanding of specific areas. Some countries want children to understand that science is constantly changing and growing. In some cases the objectives refer to knowledge and understanding of nature and its protection; to familiarization with the environment; to knowledge of the relationship between mankind and his environment, or between an organism in general and its milieu; to the principles of health, nutrition, growth and reproduction; to knowledge of how to use and transform the environment; or to an understanding of mankind itself. An understanding of the structure of international relationships and the acquisition of a global view of the world is mentioned in one case, and an understanding of the role of science in a modern world in another. Certain practical skills are mentioned: scientific skills, manual skills, work skills and work habits, and so on.

A few answers refer to the development of attitudes and interests. Encouraging a love of nature is mentioned, as is the development of an interest in studies of science and technology, and an interest in and positive attitudes towards manual and practical work.

There is hardly any area of scientific and technological curriculum content which has not been referred to in one way or another. This does not mean that all countries have stressed them, but it is obvious from the answers that all countries which teach science and technology at the primary level of education are aware of the importance of providing both a scientific way of looking at phenomena (primary goals) and knowledge, skills, understanding and attitudes (secondary goals).

In the study of *goals and objectives of person-centred change* of primary education as a whole, the integrated and all-round development of the child was found the most important one, while other possible educational developments of this kind were mentioned only in passing. The same situation was found in respect of the goals and objectives of science and technology education.

Some countries stressed the following objectives of science and technology education: pupils shall learn to live collectively in harmony with others; they shall learn to show respect for cultures, religions and laws; they shall develop an ability to take initiative. Adaptation to the environment is mentioned, as is adaptation to modern life, a changing environment, and the life which will await the pupil. The learning of behaviour related to work is emphasized in a few cases: respect for safety, welfare and hygiene; discipline at work; continuation of studies; and choice of vocation.

Secondary objectives of education refer to suitable personality development in the child: the safe-guarding of cognitive development; the all-round harmonious development of personality; the growth of capacities; and the encouragement of self-reliance, self-confidence, resourcefulness and problem-solving ability. Children shall acquire a feeling of responsibility and a solid moral foundation. They shall develop suitable attitudes towards themselves, towards reality and towards work, and they shall learn to love nature and appreciate its beauty.

All these objectives point to a need for a school system which cares for the individual child's development. This is obviously important, but a reader must ask himself whether this is enough. There are other phenomena in the school whose development deserves guidance and support: teachers and other personnel, groups of different kinds, relationships between individuals and groups and between groups, and the school as an institution. Occasional references are made to group work and co-operation but, in fact, nearly all countries have refrained from mentioning such developments.

This may seem surprising. A country cannot make use of the new developments which take place within science and technology and it cannot avoid their drawbacks except through a combined effort by groups and individuals. Working life in a modern society is based on co-operation and group efforts. Laboratories and school workshops seem suitable places to practice team work. The teacher also has an important part to play in the development of science and technology in society. He must be capable of keeping abreast of the development, of separating essential innovations from various short-lived new ideas, and of complementing new curriculum content. The faster the development, the more difficult his role and the more important the task of preparing him for his role.

Certain aspects of the development of the schools themselves and of the school system are mentioned among general goals and policy, for example adaptation to national and local conditions, and regionalization. This is important, since the school system must prepare the child for its future life. There are, however, other aspects of the development of school systems which are not mentioned at all, for example, equipping them for modern science and technology, and preparing them for changes and new roles.

Objectives of person-related change through science and technology education have been mentioned by quite a few countries, but mostly by developing and socialist ones. Few 'Western' ones report that they see science and technology education in primary school as a means to develop the child, the teacher, or the school itself.

Many philosophers, educators and politicians have looked upon education as a means to change society. The study of primary education showed many *goals and objectives of societal change:* universalization of primary education, equity of educational opportunities, social development, improvement of the living standard in the country, improved quality of life, democratization, and so on. Only a few such goals were reported in respect of science and technology education, however.

A few countries mentioned primary goals of this nature. Some referred to work processes: the application of science and technology skills in practical work, for example in manufacturing, in agriculture, in the preparation of food or in the protection of health. The preparation of workers for the labour force was mentioned in different ways: production of scientists and technicians, vocational guidance, training for team work, and so on. Protection and improvement of the environment was found an important issue: conservation of the soil, conservation of the natural milieu, environmental protection, avoidance of pollution, constructive interaction with the environment, safe-guard of resources, respect for property, and so on.

Secondary objectives are also mentioned: the learning of basic skills and productive work habits, the acquisition of proper attitudes towards work, awareness of needs for work, the acquisition of suitable consumption habits and attitudes, scientific literacy, and so forth.

These objectives always point to a need for a development staying well within the frame determined by the existing political goals of the country. Society shall be improved by the school system in such a way that it provides a better equipped labour force, but it shall not be changed in a substantial way. The latter would hardly be expected, of course: education systems serve to conserve the existing structure of society, and science and technology form a small part of the curriculum. It is of interest to note, however, that most countries mentioning objectives of social change resulting from science and technology education seem to consider only economic and manpower need. Comparatively few countries have, in fact, mentioned any objective of this kind.

That a certain objective has not been mentioned, does not prevent it from influencing curricula. It is obvious, however, that the statements of objectives made by the different Member States do not exhaust the possible benefits that could be derived from science and technology education. It is equally important that not all countries state that this education can mean an essential force to their development. This may be of less importance to developed countries with prolonged formal schooling: the students get the needed training later on or in other connections outside the school. In countries where primary education is normally terminal, the lack of a proper emphasis placed on science and technology in school may have grave effects; most young people would never receive the training they need, if the statements of goals and objectives can be counted on to reflect what is going on in school.

A general picture resulting from this analysis of goals and objectives of science and technology education is that the latter are quite conventional. As a matter of fact, most countries have reported only educational objectives. This picture may be misleading: statements of objectives may differ from the true objectives and include only the ones which are generally accepted, or the statements of the objectives of science and technology may be restricted to educational ones, while other objectives are mentioned in connection with education as a whole. That the latter is the case in many countries can be seen from the discussion of the goals and objectives of primary education above. To arrive at a more satisfactory picture of the purposes of science and technology education, it is necessary to dig deeper by studying the curricula themselves and their renewal.

A SUMMING UP

This chapter has given an introduction to the discussion of science and technology education in primary school in different countries by showing the role played by these ares of knowledge and skills in available statements of goals and objectives.

A look at the descriptions given of educational goals and objectives in different countries shows that science and technology usually play a small part. Exceptions are socialist countries with polytechnical education, since a main principle behind this system of education is to create a close link between school on the one hand, and society and the local environment on the other. In order to create such a link it has been found useful to introduce the child to activities which take place in society and to teach it skills and knowledge necessary for practical work.

To facilitate our view of the field of goals and objectives, this field was divided into four areas, which have been called goals and objectives in the following domains: explicit, ideological, cultural and political; educational; person-centred change; and societal change.

The first area is dominated by goals and objectives having to do with religion and ideology, traditional cultural values, the safe-guarding of the existing social and political system, and patriotism in a traditional sense. As far as can be judged, the value of a society which is open to scientific, technological and other innovations is nowhere explicitly mentioned among the goals.

Educational goals and objectives are in most countries dominated by the acquisition of literacy, numeracy and some civic skills. A few countries mention science and technology among valuable areas, but these subjects are never counted among the literacy skill subjects in spite of the present development of science and technology in society.

The most important goals of person-centred change are, according to available data, those of multi-faceted and all-round development of the child — intellectually, physically, aesthetically, morally and in other ways. Some countries include religious faith or political loyalty under this rubric. There are certain tendencies showing a need for a scientifically based education, namely when countries point to the goal of teaching the child to think in a scientific way — analytically, critically, with a curiosity to explain phenomena, with an ability to draw conclusions, and so on.

The goals and objectives of societal change which have been mentioned are, in the first place, those of combating illiteracy and inequity of educational opportunities. Improvement of the national economy, unity of political ideas and more profound democracy are also mentioned. Very little of what could imply a real national renewal or rapid change can be seen, but all goals and objectives are within the confines of the present national policies.

Although a scientific and technological revolution sweeps the world and scientific and technological innovations change people's lives, this is not reflected to any large extent in the statements about the goals and objectives of primary education. In many cases they seem to refer to a situation prevalent long ago. In highly developed countries, children get much of their scientific and technological education in their homes, where

Goals and objectives 95

they learn to use tools and handle everyday situations, and through the mass media, which spread information about advanced technology and scientific discoveries. In less developing countries both sources of information are completely inadequate.

Independently of whether the child gets information about certain scientific or technological phenomena or not, adequate preparation for today's world calls for a methodologically correct and systematic education in a well-prepared school in order that the need for a balanced background of knowledge and skills is met: the child shall not only acquire knowledge of individual phenomena and their functioning but also be able to place them in their perspective. Experience acquired in the home or the environment — however slight or however rich — should be used, supplemented and systematized in the school. If the home and the environment do not provide such experiences, the school must take on greater responsibilities.

Science and technology should not, of course, encroach upon religion and other culture, but it might form an important cultural element independently of the religious and political ideology of the country. Science and technology training should not replace literacy training and training in civic behaviour, but elementary scientific ways of thinking, and knowledge of skills in the handling of tools and other technological phenomena should be considered part of literacy as such and should be acquired at the same time as skills in reading, writing and arithmetic. Some countries have stressed the need for teaching children the skills of scientific thinking; such skills are necessary for other subjects in school and for everyday life, but they are often best acquired through science courses in school. Finally, a sound foundation of science and technology education among the citizens of a country is an absolute prerequisite for its economic development and ability to compete with other nations. This should always form a major aim for education, of course. Unless reponsible authorities and people in general are made to understand the role played by science and technology and the promises and threats involved in the present development, a country will not be able to develop in accordance with its potentials. This is something the present book wants to show.

Any attempt at improving the performance of education must start with an analysis of the purposes of this education: what are the primary and secondary goals and objectives. This analysis shall be connected with an analysis of the existing situation and of ways of improving the latter. The results of these analyses should be an action programme.

The data on goals presented to the thirty-ninth session of the International Conference on Education seem to indicate that some countries are still inclined to accept traditional goals for their primary education and for science and technology education in primary school without studying real possibilities. The goal data may be somewhat misleading, of course, since they do not necessarily give a correct description of the real situation. Yet, it is evident that in some cases they do not show in what way science and technology can lead to improvements in the child's way of life and in the situation of the country.

Goals must be translated into objectives, and the latter are translated into action programmes in the form of curriculum content and structure and by means of a suitable delivery system. These will be discussed in the following two chapters.

CHAPTER IV
Structure and content of primary school curricula

The discussion in the previous chapter dealt with the goals and objectives of primary education in general and of science and technology education in particular. These goals and objectives give an initial picture, often very crude and in abstract terms, of what should be taught in school — the 'intended curriculum'. They must be translatd into more or less detailed statements of what shall be taught within each subject or area during each school year and how it shall be taught — the 'planned curriculum'. This translation takes various forms according to different authorities; in many cases it is a multi-step procedure where the general curriculum outline is given by central authorities while the detailed one is prepared by local authorities, the schools or even by individual teachers.

A curriculum is a description of the way in which school authorities want the pupils to be taught in school. Heavy stress has traditionally been placed on knowledge, and the knowledge aspect still occupies a central position in most curricula. As will be discussed below, however, there is a continuing trend in many countries for other content aspects to play a larger part in the curricula: understanding, attitudes and values, the educational process, relationships between the school and the environment, and so on.

The structure and content aspects of the curricula in different countries will be discussed in the present chapter, while the following ones will deal with the delivery process, including teacher training for science and technology education, and the educational renewal process. The first part below will discuss trends in the development of curricular structures in different countries. It will be shown how the traditional discipline-oriented curriculum has been changed by the introduction of new subjects and subject matters, and how the school has become more and more integrated into its environment. The emergence of integrated curricula and integrated subjects and of polytechnical education will be discussed. The next part of the chapter will deal with the role played by science and technology in the curricula of different primary school systems. The third part will examine details of the planned curriculum, according to available documents. Finally, there is a summary with conclusions.

TRENDS IN CURRICULUM DEVELOPMENT

Traditionally, curriculum content has been divided into subjects or disciplines. For a long time, this division has followed a rather rigid pattern. Whether this pattern was been determined by practical considerations, was due to the fact that the elements of the individual subjects really did form natural units, or followed purely conventional or traditional lines is beyond the task of the present authors.

The traditional subjects are the same all over the world: religion, which in some cultures covers not only what a 'Western' educator would consider as belonging here but also other cultural manifestations; the mother tongue, often divided into reading, writing, oral expression, spelling, grammar and other disciplines; history; geography; mathematics; social sciences, often forming several subjects; natural sciences; music; drawing or arts; and so on.

During the last few decades new subjects have entered the timetable. Subjects like health, road safety and environmental protection have appeared because the physical environment has changed; others, like agriculture, gardening, practical work and technology, have been included because new roles have been attributed to the school. New content is found in several subject areas. In the case of the social sciences, natural sciences and some others, this is because the sciences themselves have changed, but in the case of foreign languages and religion it is often because these phenomena now play a different role in society.

Subjects enjoy different status. Religion, mother-tongue and mathematics have traditionally been placed high in the hierarchy, while recent innovations like technology and practical work are often less highly regarded. Undoubtedly, theoretical subjects and those which cater to the mental development of the pupils are, generally speaking, looked upon more favourably than practical subjects and those disciplines which develop the aesthetic capacity and physical dexterity of young people. The idea sometimes expressed by educational politicians that the inclusion of practical work and pre-vocational training in the curriculum should lead to manual and practical functions receiving higher status in society is not necessarily borne out by the facts.

Subjects in primary school

In many parts of the world, particularly in developing countries but also in some developed ones, the traditional subject division still exists. Although this does not necessarily mean that teaching is always discipline oriented, it shows the strength of custom.

Although many countries still prefer to divide their curricula into well-defined disciplines with traditional names, there are some variations. New subjects — or at least new names — have appeared, and a few countries have divided the curriculum into areas that differ from traditional disciplines.

The new subject most commonly encountered may be called environmental study (or studies), environmental education, environmental science (or sciences), or something similar. It may cover different areas in different circumstances, but it always serves as an introduction to the study of natural and social sciences. It is mentioned by, for example, Algeria, Bangladesh, Burundi, Cyprus, Finland, Greece, Hungary, India, Ireland, Luxembourg, Malaysia, Malta, Nepal, Rwanda, Sri Lanka, Switzerland and Zambia. A subject called 'Environmental science and applied Tongan' is found in Tonga. The *Sachunterricht* (teaching about things) subject found in Austria and the Federal Republic of Germany, the *Orienteringsämnen* (subjects for orientation) and *Naturorienterande ämnen* (subjects for orientation about nature) of the Swedish curriculum, the *Leçons des choses* (presentation of things) subject in Morocco and Turkey, the *Activités d'éveil* (awakening activities) in France and a few French-speaking countries, and the 'Contemporary studies' subject in Denmark seem to cover the same general field.

A 'nature' subject is reported by the Byelorussian SSR, China, Malta and Paraguay; a 'Nature and health education' subject by Uganda. Subjects which in some way deal with health and hygiene are fairly common. A health subject is reported by India, Malawi, Malaysia, Mexico and Sri Lanka; a science and health subject by Bahrain, Iraq and

Saudi Arabia. A hygiene subject is found in Belgium, Cameroon, Congo and Greece; a science and hygiene subject in Egypt and the Syrian Arab Republic; and health, nourishment and hygiene in Madagascar.

Technology education may be found within the framework of other subjects, but a special 'Technology and agriculture' or 'Technology and work' subject is reported by Hungary, Malawi, Mexico, Nicaragua, Poland, Sweden and Viet Nam. Work education in some form is reported by a large number of countries. Agricultural education is mentioned by Bolivia, Cameroon, Egypt, Ethiopia, Finland, Iraq, Malawi, Nicaragua, the United Republic of Tanzania and others, and gardening is found in the German Democratic Republic.

Family education and life education are subjects in Iraq, Seychelles and the United Arab Emirates; population education in Nepal; communication in Brazil and Paraguay; and leadership in Bulgaria. Vocational and pre-vocational education are found in several countries, for example, Egypt, the German Democratic Republic and Jordan. In Saudi Arabia, there are four religious subjects: Holy Quran, Quranic intonation, Islamic jurisprudence and Islamic tradition, and six Arabic language subjects: spelling and writing, reading, recitation and songs, dictation, handwriting, and composition, in addition to other subjects commonly found in primary school curricula.

This list of comparatively new subjects is far from exhaustive, but it does illustrate the way in which curricula are developing. Most of these new subjects have a scientific or technological inclination; they are often simply parts of a previous science course which have become subjects in their own right. It is quite obvious that such a development need not necessarily mean that anything new has been added to the curriculum, nor that science and technology have been given stronger emphasis.

There are quite a few countries where the trend has gone even further and where a complete or partial revision of the distribution of the curriculum into subjects has been made. Let us examine a few examples of this development:

In Madagascar there is a series of subjects or subject areas: *Fiarahamonina* (history, geography, civics; learning to live; socio-cultural and political education); *connaissances usuelles* (common knowledge; health, nutrition, hygiene, elementary science in practical life); practical and artistic education (agriculture, practical work, increasing awareness, socialist education); mother-tongue, mathematics, French and physical education.

In Thailand the curriculum content is divided into four subject areas: skill subjects (Thai language, mathematics); life experience (health, nutrition, general family education, social and community life); character development (moral education, arts, physical education); and work-oriented subjects (handicrafts, agriculture, etc.). In addition, pupils may take English.

In the first three grades in primary school in Paraguay, education is given within three areas: social life and communication; nature, health and work; and mathematics. In the following three grades, the first two areas are broken up into their constituent parts. Each subject area is further divided into a number of units.

Complete or almost complete avoidance of subject divisions is also reported. Tunisia mentions the subject 'Arabic', which is the only one in grades 1 to 3. French is added in grades 4 to 7 and, in some schools, manual work in grades 5 to 8. 'Arabic' thus covers Arabic language, mathematics, awakening activities, civic and religious education, artistic education and other areas. In several countries, for example, Colombia and Mexico, education is completely integrated in the lower grades.

Structure of curricula

The preceding text has already pointed to a tendency in some countries to combine subjects or create new, rather comprehensive ones, although the vast majority still retain traditional subjects. A curriculum split up into a number of separate subjects does not necessarily mean that they are taught in isolation. On the contrary, on many occasions it has been stated that some kind of integration shall take place; elsewhere, even at the primary-school level, the best procedure is considered to regard subjects as independent disciplines to be taught separately. On the other hand, combining subjects into one or several 'integrated' areas does not necessarily mean that teaching practice follows the new directives. It is quite conceivable that the old subjects survive within the new system. Two trends have, however, changed the picture of a subject-oriented primary-school curriculum: integrated curricula and polytechnical education.

If the different subjects are taught in isolation, stress is placed on the building up of a theoretical structure such that the characteristics and internal logics of each discipline are emphasized. An aim is to develop the student's understanding of the principles behind the subject, such as models, theory and application of theory. The content of the subject matter of each discipline forms, in principle, an integrated whole where each new part or unit builds on previously studied units, as in mathematics.

Such theoretical approaches call for a high level of maturity on the part of the student. Therefore, discipline-oriented curricula are found primarily at advanced levels of schooling. Subjects like physics, chemistry and biology usually make their appearance towards the end of primary school or later. At the upper-secondary level, a discipline-oriented curriculum is the rule with few exceptions.

If schooling at one level is destined to prepare mainly for schooling at a subsequent level, and if the final aim is profound scientific understanding, discipline-oriented teaching seems inevitable. The late Professor Max Beberman, one of the fathers of 'new maths', once said that you have to teach 1,000 pupils mathematics to find one mathematician. However, the idea of education forming a closed field, 'education in a glass bowl', is no longer generally accepted. A prime objective of education is to prepare the child for practical life and work. In most cases, particularly in developing countries, children enrol in and attend primary school to get something useful out of it, without intending to continue at higher levels. In such cases discipline-oriented approaches at the primary level are less suitable than child-centred ones. This is one of the reasons for innovations like integrated curricula and polytechnical education.

The integrated curriculum takes the child's world as its starting point. It is organized in accordance with problems young people face or are likely to face when they leave school and lead adult lives. Such problems may have to do with their body — health, nutrition, growth, etc.; others with their ways of living — further education, training, jobs, leisure time, relationships with the law, etc.; or with their relations with other people — sex, family life, clubs, sports, etc.; and so on. An aim of formal education is to enable young people to solve such problems. For this reason, they shall be given an all-round, multi-faceted education which is also problem centred. The internal organization of each discipline will then be considered less important than its relationships with and value for mankind. The content of the education given shall be appropriate to the age, experience and development of the child, but also to the problems he/she will face. At first, the teacher may draw from experiences already acquired, but as problems become more and more complex, special information and skills must be provided. Although children are not alike and their problems may differ, the necessity of providing solutions to their future problems means that a common core curriculum should be devised and taught in school.

A recent realization is that early differentiation into several disciplines is not necessarily the best way for the school to introduce the child to the complexities of modern society. The child should learn to handle life problems, and the curriculum should be constructed accordingly. Real-life problems are rarely confined to subject areas but call for information from several fields.

A differentiation can sometimes be made between integrated subjects, that is, those which bring together several disciplines, and integrated curricula, which combine a major part of the whole curriculum. This differentiation is often unimportant and will not be pursued here. Integrated curricula can certainly be traced back in history, but the application of the idea on a large scale seems a recent phenomenon in most countries. Most modern educators, however, seem to agree that primary school must help the pupil to get an age-adapted insight into nature, work and society, and that the best way of doing this is to start from the familiar environment, gradually widening the circle. In this way, the pupil will become familiar with phenomena and can come to understand intricate relationships. Only then will he/she become mature enough to grasp the theoretical structure represented by a particular discipline. It may be that this process is facilitated if the curriculum is undifferentiated, and if multi-disciplinary themes or projects are introduced.

The modern way of structuring the curriculum — polytechnical education — is based on similar principles. In choosing areas for a common curriculum, attention is paid to the needs of society. The relationship between the world of work and social activity on the one hand, and what is taught in school on the other, is of paramount importance when choosing curriculum content. In teaching particular content, the teacher makes reference to the usefulness and social implications of what is being learned, whether a practical skill or a theoretical principle.

This approach is intimately related to educational ideas accepted by socialist countries in Eastern Europe and other parts of the world, but it is discussed and tends to influence educational development in other countries as well. It means an attempt to form a bridge between theory and practice, academic education and vocational training, school and life. Courses taught in school are directly connected with social reality and the labour market in the society in question. In the first few grades, pupils typically acquire basic skills in reading, writing and arithmetic, and gain an age-adapted insight into nature, work and society within often rather traditional subjects. Instruction is linked with socially useful work. Later on, the pupils become familiar with social life, work, science, technology and culture, and are more and more integrated into practical work. Finally, work training is given. Countries with polytechnical education may also use integrated curricula.

Curriculum approaches in different countries

As was mentioned above, most countries still divide the school day into periods for the study of traditional subjects like mother-tongue and mathematics. Whether this means that the teaching is discipline oriented and that stress is placed on the internal structure of the subjects can only be determined by a scrutiny of the ways in which subjects are actually presented to the pupils. In most cases there is ample evidence that teachers follow traditional ways of teaching subjects, at least in the upper grades. Even in countries with subject-oriented curricula, some integration between subjects is usually sought in the first few grades.

A subject-oriented curriculum is found in Nicaragua. In the first grade, science and technology form components within the subject Spanish. In the following grades, however, they form separate subjects: natural sciences and agriculture from grade 2; technical-industrial education from grade 5. In Gabon, Kuwait and the United Arab Emirates, the traditional curriculum includes separate mathematics and science subjects; and in Egypt it has been modernized to include health, agriculture and science (grades 5 to 6) with practical application and co-operation with other subjects, but it is still subject oriented.

In Paraguay, integration between subjects is the method used during the first three years of primary school. As mentioned, there are three major subject areas. During the second part of primary school, these areas are divided into their constituent parts and the teaching becomes subject oriented. In Belgium, there are separate subjects for certain curriculum areas like mathematics from grade 1. Until grade 4, the teaching of geography, history and science is conducted in a global way as an exploration and conquest of the environment.

The introduction of environmental nature study, or a similar subject, means that the boundaries between traditional subjects have been crossed and that the first large step towards an integrated approach has been taken. This is also expressed by some countries. The Tonga subject 'Environmental science and applied Tongan' means an attempt to integrate various sciences under a major topic. Sri Lanka states that, at the lower primary stage, science forms a part of an integrated environmental studies curriculum. Switzerland mentions that, in the majority of cantons, there are several subject areas, one of which is 'Knowledge of the environment'. In Denmark, the subject 'Contemporary science' integrates history, geography and biology. In the Hungarian school timetable, 'environment' includes science from grade 1, technology from grade 2, geography from grade 6 and chemistry from grade 7. It is said that science and technology teaching takes place within the framework of traditional subjects, but the subject matters are interrelated by their content and function, and not isolated from each other.

More sweeping integration of the curriculum also exists. Thailand's four major subject areas have been mentioned before; within the area 'Life experience' they allow an integrated approach to topics like health education, nutrition, education about the family, and social and community education. In Tunisia, the subject Arabic covers everything except French and manual work. It is stated that this educational innovation enables the pupil to accept his role in his milieu; to acquire an opening to modern technological culture; to strengthen moral, civic and social education; and to improve educational methods. In Chile, the curriculum is divided into several areas with traditional activities: verbal expression, numerical expression, experience (society, geography, history, science and religion) and so on. This allows the integration of social science and natural science subjects into a whole. In Colombia, three ways of integration between subjects are sought: within each curriculum area, between different areas, and between the educational process and the community. There is also an attempt to provide continuity between different grades and levels.

The idea of an integrated curriculum seems commonly accepted among Western European countries. In the Federal Republic of Germany and Austria, the major subject *Sachunterricht* integrates social studies, biology, physics and chemistry with other subjects and provides basic knowledge. Bulgaria mentions an integrated curriculum: during the first three years the school system aims at an integrated development of the child's personality based on language skills, an introduction to quantitative relationships and the nature of society, and a formation of basic values. The educational content is organized into global themes with integration of all disciplines. France reports that the primary school curriculum devotes seven hours per week to the subject called *Activités d'éveil*, which incorporates social studies, environmental studies, science and so on. Only general objectives and guidelines exist within these areas, however; teachers are given a great deal of freedom. As mentioned above, Sweden has a similar integrated subject, *Orienteringsämnen*, which deals with social studies and science. Other European countries also report efforts at integration.

In Malaysia, mathematics forms a separate subject throughout primary school but, in the new curriculum, science is taught as an integrated subject. In the first phase, grades 1 to 3, it is absorbed into the language programme, but in the second phase it is integrated into the area 'Man and his environment'. Morocco integrates science and technology courses with other disciplines. Algeria uses an interdisciplinary perspective: mathematics, scientific and technological introduction along with other activities. However, certain courses seem to form sub-disciplines within the general area. Uganda has an integrated subject which includes agriculture, various natural sciences and technology. Mathematics is taught separately. In the United States, science education is generally taught in an integrated approach according to themes, and in mathematics there is also a general course broken down into themes. In Bangladesh, the subject 'Environmental studies' comprises health and population education, agriculture and crafts.

In some countries the system varies due to the fact that, within the framework of general rules, the schools themselves determine the curriculum. Australia reports that science courses taught in primary school range from teacher-designed programmes of work to the adoption of State and overseas curricula. Science may be taught as a separate subject, but it is usually taught in an integrated way. The United Kingdom states that the responsibility for the selection and provision of scientific experience lies with the individual teacher and that courses vary enormously. Courses are not always taught with separate disciplines, however, but often as parts of integrated programmes.

A summing up

It has already been stated that education systems all over the world have been built up on the basis of a model which has its roots in the traditional European system. This also applies to the division of the curriculum into subjects. In developing countries they are largely the same as in developed ones. There are exceptions to this rule: countries like Chile, Colombia, Thailand, Tonga and Tunisia have devised interesting curricula which may be suited to their particular situations. In general, however, the impetus for educational renewal seems to originate in developed countries. Polytechnical education was developed in Eastern Europe but has now been adopted by many developing countries. The integrated approach has its roots in Western European countries like the United Kingdom and Sweden, but has spread both in the form of an integrated subject like 'environmental studies' and in the form of completely integrated teaching.

Science and technology are at the centre of renewal in curriculum content and structure. When studied as specific subjects, science is usually found only at the upper levels of primary school. The structure of the natural sciences is usually such that a high level of maturity is required for pupils to grasp the meaning; at the same time, knowledge of science is necessary to comprehend problems in the life of a young person. The solution to this dilemma has been an integration of science with social sciences and other subjects. Technology, on the other hand, is not considered to be as difficult.

THE PLACE OF SCIENCE AND TECHNOLOGY IN PRIMARY SCHOOL CURRICULA

School timetables from around the world reveal that science is taught under different names. Physics, chemistry and biology are obvious candidates. Geography, health, nutrition, hygiene and similar subjects mentioned by different countries may also include science elements. Environmental and nature subjects provide an introduction to science. Technology may form a separate subject, but it is more common to include technology elements in other subjects. In the following, a brief overview will be given of science and technology subjects in primary schools in different countries.

Science education

Many countries report *traditional science subjects* from an early grade. Ethiopia mentions science in all grades but also states that teaching is integrated. Lower grades get more training in scientific skills, while in the upper grades there are more scientific facts and practical activities related to the immediate environment. In Nigeria, the curriculum includes the subjects of science, health and physical education. The former is taught as an integrated subject, and a core curriculum for science education has been developed. Malawi has thirteen primary school subjects including health education (all grades), science (all grades) and geography (all grades). In Congo, there is hygiene from grade 3, geography and science from grade 5. Cameroon has a primary school timetable which includes moral education and hygiene (grades 1 to 4), science (3 to 6), and history and geography (4 to 6). The United Republic of Tanzania teaches the subject general science in grades 3 to 7 and domestic science (hygiene and home sciences, etc.) in grades 1 to 7. The teaching is always according to separate disciplines.

Japan, Seychelles and Viet Nam teach science from grade 1; the second-mentioned country reports that the curriculum is integrated during the first three years. In the Republic of Korea, science is introduced in grade 2, and in Pakistan there is science in all grades, and health and physical education in grades 3 to 6. Most Arab States report science in all grades. In Iraq there is 'Life education' in grades 1 to 3 and 'Science and health' in grades 4 to 6. In Algeria, the subject 'Studies of the environment' appears from grade 3.

Mexico reports an integrated curriculum in grades 1 and 2. Some science is taught in grades 3 to 6, as is health education. Bolivia and Guyana mention science and health education in all grades. In Bahamas, there is science in all grades; it includes health education, environmental studies and nature study. In Argentina and Jamaica, there is integrated science teaching in all grades; and in Nicaragua, the same is true in grades 2 to 6. Peru reports natural sciences in all grades but also geography in grades 5 to 6. Cuba mentions science from the third grade. In Colombia the existing ('old') curriculum includes science throughout primary education, while the 'new' one implies an introduction of science and health in grades 4 to 5 after integrated teaching in the first three grades.

An *environmental subject* gives an introduction to both science and social studies, but in most countries where this happens science subjects re-appear in the upper grades.

In Zambia there is environmental science in all grades. It combines biology, chemistry and physics. In Rwanda, courses include both environmental studies and science in all grades, and in Burundi there is environmental study from grade 1. Mauritius reports that science is integrated with social studies and home economics under 'Environmental studies'.

Nepal states that several science areas are covered: 'Fundamental knowledge in science and health', 'Environmental education' and 'Population education'. No science is taught in the first three grades, and it is always taught as a separate subject. India mentions the subject 'Environmental study' (grades 1 and 2), 'Health education and games' (grades 1 to 8) and general science (grades 3 to 5). Bangladesh includes environmental science (all grades) in her primary school curriculum; and Malaysia includes the integrated subject 'Man and environment' (from grade 4) in her new one. In the first three grades, science is included in the language programme. The existing five-year primary school in Turkey teaches the subject *Leçons des choses* in grades 1 to 3 and science from grade 4. In Sri Lanka, environmental studies covers health habits in grades 1 to 3 and science in grades 4 and 5. Scientific processes are said to play a major role in the learning process at the primary level.

In the majority of Swiss cantons there is the study area 'Knowledge of the environment', which covers geography, history, science, etc. Science forms a small part of it. In the first three years, children are made aware of science during environmental study, but from the fourth grade there is some specific training. Although individual subjects still exist, teaching often adopts interdisciplinary themes. Ireland reports environmental studies for the age range 4 to 8 years in which children learn social and environmental facts. Then the subject area 'Social and environmental studies' appears, and from age 10 subjects like history, civics, geography and elementary science are found. There is no prescriptive programme, merely suggestions and advice on how to exploit the children's environment within the context of an integrated curriculum.

Malta states that from grade 4, the curriculum expands to include geography, nature study and environment. Greece mentions the subject *Etude du milieu* (study of the environment) in grades 1 and 2, and the subjects physics, chemistry with hygiene, and geography in grades 3 to 6. Finland has environmental studies in grades 1 to 4, natural history and geography in grades 3 to 9, and physics and chemistry in grades 7 to 9. In Sweden there is, as previously mentioned, an *Orienteringsämnen* in grades 1 to 3. It splits up into social science and a pure science in grades 4 to 6.

In Poland, children study the social and natural environment in grades 1 to 3, biology and hygiene in grades 4 to 8, geography in grades 4 to 8, physics in grades 6 to 8 and chemistry in grades 7 to 8. Czechoslovakia reports 'introductory lessons' in grades 1 to 2, and then history and geography (grades 3 to 4), geography (grades 5 to 8), natural history (grades 5 to 8), physics (grades 6 to 8), and chemistry (grades 7 to 8). The German Democratic Republic gives age-adapted insights into nature, work and society in grades 1 to 2; then biology and geography are introduced from grade 5 and physics from grade 6. In Bulgaria there is 'Study of the native country' from grade 1. This involves many scientific areas. The natural sciences, physics and chemistry become independent from grade 3, but the scientific initiation is said to have been covered by other subjects. The method used is to integrate areas from several subjects without diminishing the autonomy of the disciplines. In Hungary, pupils are taught environment in grades 1 to 5, geography in grades 6 to 8, physics in grades 6 to 8, biology in grades 6 to 8 and chemistry in grades 7 to 8.

Nature subjects are directly concerned with science. Uganda reports nine subjects in primary school, of which nature and health education occur in grades 1 to 2 and science in grades 3 to 7. In China, there are courses on nature in grades 3 to 5 and geography in grade 4. The 'Nature' area is a synthesis of basic physics, chemistry, astronomy, geography, biology, etc. In Romania, the science subjects are 'Knowledge of nature and home' (grades 2 to 4), and geography (grades 3 to 4). In the USSR, there is a nature study subject in the lower grades, followed by systematic study of individual disciplines in the following ones.

An *integrated curriculum* with content covering science together with other subjects is a common phenomenon.

In Thailand the 'Life experience' area covers health education, nutrition, etc. In Turkey, the new, only partly introduced basic education of eight years includes science in all grades. During the second and third year there is an introduction to science education through the subject *Leçons des choses*, which integrates science and social study. During the fourth and fifth year, more profound science education is given in the subject *Connaissances scientifiques* (scientific knowledge). In Tunisia, the subject Arabic also includes science; in France the integrated area *Activités d'éveil* brings up scientific and other content matter. In Austria and the Federal Republic of Germany, the subject *Sachunterricht* employs scientific methods and knowledge to clarify problems, questions and phenomena within the child's range of interest; this is evidently also the main idea behind curriculum integration itself. *Sachunterricht* thus gives an introduction to selected scientific disciplines. In Denmark, there is the subject 'Contemporary studies' in grades 3 to 5 and physics and chemistry in grade 7.

In Paraguay, as already noted, the integrated approach in the first three grades subsequently becomes subject-oriented. In Chile as well, there is a common science programme in grades 1 to 4, teaching becoming increasingly specialized. In Belgium, natural sciences appear from the fifth year, no distinction being made between biology, physics and chemistry. Geography is taught together with history. No premature differentiation is sought.

During the first few years of schooling in Brazil, biology, physics and mathematics are treated as activities associated with social studies and communication. Later on, the sciences are treated as areas formed by the integration of content in the subjects biology, physics and chemistry. In Mauritius, science is integrated with social studies and home economics under environmental studies. In Cyprus, there is a separate science subject, but efforts are made to integrate it with other subjects like mathematics, geography and study of the environment.

Gabon and Morocco report that the curriculum leaves little room for science, and the Central African Republic has a discipline-oriented curriculum with a few hours devoted to science. In the latter case, however, it is reported that in the new curriculum, the subjects *Science d'observation* and *Etude du milieu* (observation science and environmental studies) have been combined into a new subject area to be taught from grade 1. In Australia, the United Kingdom and the United States, there are no fixed timetables, and the time set aside for natural sciences varies. In Australia, science may be taught separately as a subject but, in accordance with the established processes of learning, subject matters from science areas are usually taught in an integrated thematic form. In the United Kingdom, the responsibility for the selection and provision of scientific experience lies with the individual teacher, and courses vary enormously. The latter are not always taught as disciplines but often as parts of integrated work. The Netherlands reports that the content of science education is determined by the schools. In most cases, only biology and selected areas of other disciplines are taught. Traditionally, such subjects are taught separately, but emphasis has shifted to integrated courses in which the teaching is often thematic.

Technology education
Some kind of education with technological content is reported by most countries. Although there is a distinct technology subject in a few countries, technology usually forms an integrated or discrete part of some other subject or activity. The size of this part is usually hard to determine, but it can be assumed that it is rarely large.

A separate *technology subject*, often quite important, is characteristic of primary education in socialist countries. In Hungary, it occupies 6 per cent of the total timetable in the first eight grades. In Poland, two hours per week are devoted to 'Work and technology' in all eight grades. Romania reports two such periods per week in grades 5 and 6, one in grades 7 and 8. In the German Democratic Republic there is 'Polytechnical training' in grades 7 and 8 for four hours per week. Nicaragua mentions the subject 'Technical-industrial orientation' with two periods per week in each of grades 5 and 6. In Algeria, there is 'Technology and applied science' with one-and-a-half periods per week in grades 1 to 3, one in grades 4 to 6, and three in grades 7 to 9. A separate technology subject is also found in the Netherlands, where special classes have been organized in a number of schools to make children familiar with technological development and teach them various techniques. Sweden has technology education in all grades.

In Chile there is a subject called 'Technical-manual education', and in Mexico one called 'Technological education' which occupies about 5 per cent of the time in primary school. Malawi has 'Craft and technology' in grades 3 to 8, and the United Arab Emirates home economics (for girls) or technology (for boys) in grades 4 to 6. The new school in Turkey is reported to have technology education of between four and eight periods per week in grades 4 to 8.

Technology education is reported to form part of the training in a traditional *theoretical subject* by a large number of countries. In most cases this subject was science, reported by, for example, Argentina, Bahamas, Brazil, Central African Republic, Chile, Cyprus, Iraq, Jamaica, Jordan, Mexico, Nepal, Nigeria, Paraguay, Peru, Senegal, Seychelles, Spain, Turkey (new school system), Uganda and Zambia. In the Central African Republic and Congo, however, the subject is called 'Observation science'; in Spain, 'Social and natural experience'. In Uganda, it is said to be science and its application to crop and animal husbandry, sanitation and home construction; in Iraq and Turkey it is science and health; and in China and Paraguay a nature subject. An environmental subject, said to include technological training, is reported by the Central African Republic, Finland, the German Democratic Republic, Hungary, Iraq, Switzerland and Thailand; in Cyprus there is a 'Life experience' subject.

Natural and social sciences are sometimes found to provide technological education. Physics is mentioned by the German Democratic Republic and Ireland; chemistry by Hungary; biology by Finland and Hungary; and geography by the Central African Republic, China, Cyprus, Hungary, Japan and the United Republic of Tanzania. That science subjects contain technological components is natural; in many cases technology education is defined as practical training in the use of what is learned in science education. Some countries, however, report that technology forms part of a nonscientific subject: history (Congo), social studies (Bahamas, Finland, Jamaica) or language arts (Bahamas, Madagascar).

Technology is obviously closely related to *arts and crafts subjects* with the common denominator that they all imply the use of hands. The subject arts and crafts is mentioned by Austria, Denmark, Finland, Guinea, Jamaica, Japan, Mauritius, Pakistan, Paraguay, Tonga, Uganda, the United Republic of Tanzania and Zambia; handicraft or a similarly named subject by Cameroon, Denmark, Guinea, Nigeria, the United Kingdom and Zambia; drawing by Congo and Pakistan; home making by Japan; home economics by the United Kingdom and Paraguay; domestic science by the United Republic of Tanzania; and home construction by Uganda.

Many countries find that different *work or vocational subjects* provide technology education. Manual work or a similarly named subject is reported to do so by the German Democratic Republic, Madagascar, Nigeria, Senegal, Switzerland and Turkey (new school system); vocational education by Jordan; productive work by Algeria, Congo and Tunisia; gardening by the German Democratic Republic and Senegal; and some other work-related subject by Bulgaria, China, Tonga and Zambia. Bulgaria states that 'Work and creativity' includes, above all, elements of technological education and an introduction to productive work.

Agriculture, animal husbandry or a similar subject with technological content is reported by Bahamas, Bangladesh, Cameroon, Guinea, Nicaragua, Nigeria, Paraguay and the United Republic of Tanzania. Among other subjects or areas mentioned by different countries are computer education (United States) and laboratory training (Cuba). Reference has already been made to the *Sachunterricht* subject in the Federal Republic of Germany, to the *Activités d'éveil* in France and other countries, and to the *Leçons des choses* in Turkey, all of which provide technology education. A few countries report that all subjects or areas have a technological component: Colombia, Cuba, the United Arab Emirates and others. Little or no technology education is said to be given in primary schools in Australia, Malta and Malaysia.

Apart from the cases where technology forms a subject of its own with a specified number of hours per week in the timetable, it can be taken for granted that limited technology education is given. All data indicate that technology forms a comparatively small part of the subjects with which it is connected, and teachers would be likely to stress the traditional content of the latter. Furthermore, many of these subjects are themselves given limited time.

Another source of data on technology education in the world is provided by the Unesco study *Technology education as part of general education*[84] which was based on a survey of thirty-seven countries. It showed the same tendency as the present study: a specific technology subject is found in a few cases, but technology is reported as part of arts and crafts, domestic science, home economics, shorthand, agriculture, industrial, commercial, natural sciences and other programmes.

The role of science and technology

The exact proportions of the curricula which are devoted to science and technology cannot be determined. An attempt at studying timetables in different countries shows that between zero and 20 per cent of the time is devoted to science, the highest figure being obtained only by assuming that *all* environmental study is scientific. An average would be closer to 5 than 10 per cent. With regard to technology education, few non-socialist countries devote more than one or two hours per week to the subject; in most cases, no time at all except some necessary measurement in a science subject. This can be compared with the time devoted to mother tongue, about 30 per cent of the timetable; mathematics, approximately 20 per cent and in some cases over 30; social studies, almost 10 per cent, in some cases almost 20; and religion or moral education, between 5 and 10 per cent in the countries where such education is taught. The impression gained from the study of the goals and objectives of primary education are confirmed: priorities are given to the study of reading, writing, arithmetic, and some social and humanistic areas, not to science and technology training.

There are a few other findings which should be considered in an evaluation of science and technology education. The first is that such education is found mainly in the upper grades of primary school. When science is taught in the lower grades, it is nearly always as part of an environmental subject or occasionally an integrated science subject. Another fact is that considerably more science and technology is taught in schools in developed countries than in developing ones. Reasons for this may be connected with the higher costs of teaching these subjects, a lack of teachers with suitable backgrounds, or the priority given to other teaching, for example, literacy training or religious instruction.

This should be considered in relation to other facts that have been discussed in this volume. In developed countries, enrolment at both primary school and lower secondary school is usually complete; science and technology instruction are provided at both. Most young people also attend upper secondary school. Since they get scientific and technological training at the higher levels, there is no absolute need for such training in the primary grades. In developing countries the situation is quite different. Fewer young people attend secondary school. In addition, the environment is poor in scientific and technological stimulation. Therefore, primary education should play a much more important role.

It has already been stated that it is difficult to teach science at an early age if the instruction is meant to provide insights into the theoretical logic and structure of each discipline; this calls for mature students. However, the aim of science education has changed in most countries. At least in the lower grades, it is usual to study natural phenomena and their importance and to learn to understand the relationship between nature and Man. Only later are theoretical aspects considered.

There is no generally accepted definition of technology education, and the content varies according to different circumstances. In some cases it means learning to use simple tools, in others the practical application of scientific knowledge, and in still others it may consist of advanced skills like computer programming. During the thirty-ninth session of the International Conference on Education, it was pointed out that it is necessary for children to learn how to use simple tools in school. In developed countries the children's homes provide some such training, and therefore more advanced technology can be introduced at an earlier stage. For this reason, children in such countries are placed in a favourable situation.

THE CONTENT OF SCIENCE AND TECHNOLOGY EDUCATION

To present detailed content descriptions of science and technology education in all member States would be neither possible nor very useful. It would be impossible not only because it would be hard to obtain such descriptions but also because detailed, nationally accepted curricula do not exist in countries where curriculum preparation is decentralized or where schools and teachers are given a great deal of freedom to prepare their own lessons. It would be of little use, because the individual content elements form a selection of possible items chosen more or less systematically on the basis of perceived national and local needs, traditions and other factors, which means that countries in other situations would have to make a different choice. Furthermore, the multitude of items would make it hard to determine essential trends. The important point in this discussion is not what elements are included but how they are organized in a curriculum designed to meet modern demands.

In the following text, the teaching of science and technology in a few countries will be discussed. In the case of science, only countries with an integrated science course or integrated curricula will be covered. The approach in these cases is usually in the form of interdisciplinary themes or units which have been chosen on the basis of problems appropriate to the pupils' milieu or in their life situations. This is an approach which deviates from traditional curriculum designs, and a discussion might be profitable.

Content of science education

This discussion will start with a few examples of the content of science education in countries where this subject is taught in an integratd way from the lower grades of primary school. An attempt will also be made to show how integrated teaching turns into subject-oriented teaching. Finally, some examples of polytechnical education will be mentioned.

Paraguay presents a typical example of the way in which science education is initiated in a country with an integrated curriculum. The subject 'nature, health and work' is found in grades 1 to 3 and corresponds closely to the environmental studies and similar subjects in other countries. It is divided into six themes: 'Child and environment' with scientific and other content; 'Child and work' with mainly technological and work education; 'Child and health' with mainly scientific content; 'Child and family' with scientific and other content; 'Child and space' with mainly scientific content; and 'Child and play' without scientific or technological content.

The first theme concerns the interpretation of the phenomena around the child and discusses interrelationships between living beings and their environment. The approach is ecological, and scientific thinking is encouraged. The third theme aims to help the child develop habits which create physical and mental balance and to acquaint the child with health services as a means to protect and improve health. The theme about child and family considers the latter as a social nucleus in which the child is born, learns different things and acquires habits. The space theme is intended to teach basic concepts about earth and space.

In the following grades — 4 to 6 — there are individual subjects: science, health and work. Science is integrated with the following units: 'Living beings and their interactions with their environment', 'Beings in nature and their organization', 'Man and the changing world', 'Man and balance in nature' and 'Man and space'. Health covers the following units: 'Man and his family', 'Man and health', 'Man and sports and rhythmical movements' and 'Man and protection against illness'. The integrated approach during the first three years thus turns into a systematic one during the following period.

In Colombia, the integrated curriculum, which covers almost all subjects under one heading, is divided into eight basic areas of knowledge. Among these are 'Science and health' and 'Technology'. The teaching of science is built round a conceptual model of natural sciences and health. It can be described by means of the model given in Figure 4.

In the content description, the following key words, corresponding to areas or themes, appear: 'Structure of the universe'; 'Interaction—change'; 'Interaction-conservation'; and 'Applicability'. The following and many other elements are taught in the different grades: *Grade 1:* living and inanimate things; man as a living system; living things in the environment; sun, earth, moon and universe; the rotation of the earth and the creation of day and night; the week. *Grade 2:* the variety of living and inanimate things in the environment and relations between them; the activity and inactivity of living things in nature; transport, energy and work; length, mass and weight; duration; different manifestations of matter — solids, fluids, gases; earth movements. *Grade 3:* families of beings; substances in the form of mixtures and combinations; energy as a prerequisite for the functioning of plants, animals and machines; light; the moon as a satellite; measurement. In the following grades, science is taught within the framework of the subject 'Natural sciences and health'. The teaching is integrated within the subject boundaries in accordance with the general principles of the system.

Turkey is also introducing a new primary school system where scientific facts will be presented within the subject *Connaissances scientifiques*, which is taught in all grades. The programme covers the following areas: earth; the sky; natural resources; matter and energy; our body and our health; technological progress; and space.

In Iraq, science education in grades 1 to 3 takes place within the subject 'Life education'. In grade 1, teaching concerns the school and social habits; the family, good manners and responsibility; the house and its function; the neighbourhood, transportation and services; and the local market. In grade 2, the syllabus brings up the different seasons — autumn, winter, spring and summer — and the teaching concerns climate, clothes, ways of living, illnesses, natural phenomena and other matters related to the seasons. The third grade brings up for discussion the human body and its care, nutrition, health care, farming, industries, social services, transportation and the country.

In grades 4 to 6, science is taught within the subject 'Science and health education', which aims to give the pupil essential information and skills, such as how to use simple tools and instruments as well as the ability to read and understand instructions. In the fourth grade, the syllabus includes elements of self-protection, water in our life, animals and their uses, plants, magnetism, etc. In the fifth grade it covers air in our life, crops, insects, energy, sound and light. The sixth grade deals with our body and its care, diseases, first aid, medicine, local industry, electricity and the planets.

Among the eight programmes in primary school in Mexico, one is called 'Natural sciences' and another 'Technological education'. During the six grades, science educa-

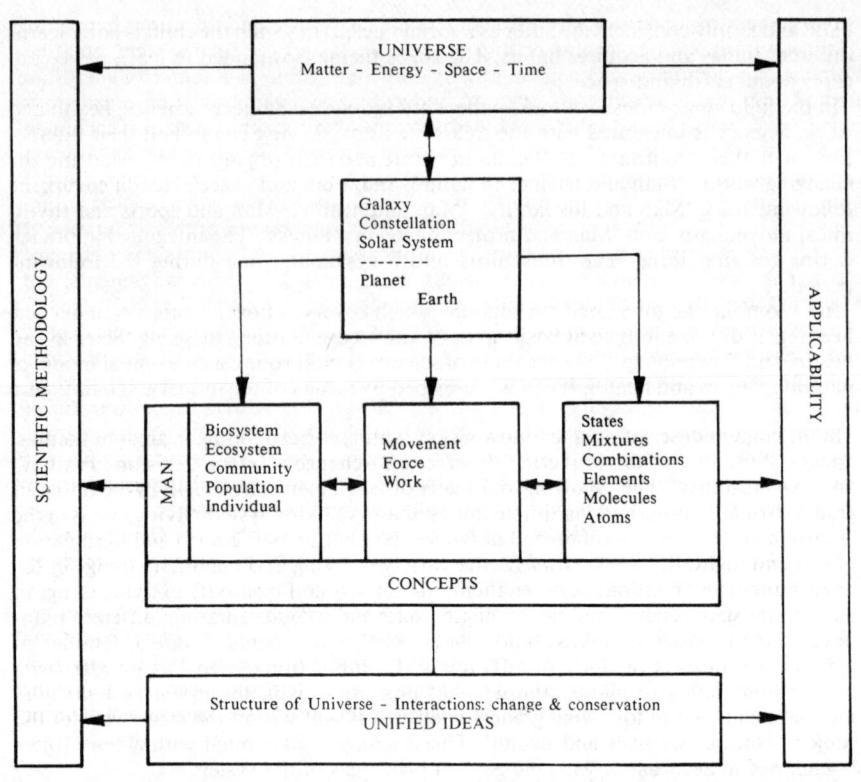

```
                    ┌─────────────────────────┐
                    │        UNIVERSE         │
                    │ Matter - Energy - Space - Time │
                    └─────────────────────────┘
                              ↕
                    ┌─────────────────┐
                    │ Galaxy          │
                    │ Constellation   │
                    │ Solar System    │
                    │ - - - - - - -   │
                    │ Planet          │
                    │      Earth      │
                    └─────────────────┘
```

FIGURE 4. Conceptual structure of the natural sciences and health, according to a curriculum document from Colombia

(Diagram elements: SCIENTIFIC METHODOLOGY ↔ [MAN | Biosystem, Ecosystem, Community, Population, Individual] ↔ [Movement, Force, Work] ↔ [States, Mixtures, Combinations, Elements, Molecules, Atoms] ↔ APPLICABILITY; CONCEPTS; Structure of Universe - Interactions: change & conservation — UNIFIED IDEAS)

tion covers four major themes: living beings; the environment; matter and energy; and the stars and outer space. The aim is to create a scientific attitude in the children which allows them to see science as an evolutive process in which phenomena are studied in a systematic manner. The first two grades have an integrated curriculum. The child learns simple concepts, processes, operations, etc., and carries out experiments in an integrated way. In grades 3 to 6 there is a specific science subject, as mentioned, but teaching follows the four themes.

In grade 3, the child is introduced in a simple way to the study of natural sciences, to their principles and to the general procedures used. The properties of water and the phenomena which change nature are studied to bring out the concept of causality and an understanding that matter is transformed, not created or destroyed. The characteristics of jungle, forest and desert are studied, as is the human body, its organs and functions. In the fourth grade, the pupil studies the concept of energy in different forms and is introduced to the fact that matter exists as solids, fluids or gases. The hydrological cycle, the participation of Man in social change, food and digestion, space, spatial and celestial phenomena, and the human body are presented. In the fifth grade the stress is on practising the scientific method of observation. The child is alerted to the relations

Structure and content of primary school curricula

between the components of a terrestial or aquatic ecosystem, modifications happening in the environment, the characteristics of the planets and the solar system, etc. Phenomena like electricity, magnetism, gravity and combustion as a source of energy are treated, and finally the child is made to understand the reactions of an organism to phenomena like light and sound, and the functions of the nervous system. Teaching in grade six brings up the reproductive apparatus in Man and the cell as the fundamental unit in the reproductive process. It also mentions human nutrition, a balanced diet, animal and vegetable food, and nourishment. Evolution is discussed, as well as animal behaviour, the capacity of animals to adapt and the importance of this adaptation to survival. The ecosystems of the American continents are treated, as are surface changes, the formation of mountains, the distribution of stars and their movements, etc. Machines and work of different kinds are discussed.

In the primary education course in Cyprus, efforts are made to let the pupils explore and understand natural and man-made environments and familiarize them with scientific approaches. This takes place first within the subject 'Environmental studies', which also contains a section on social studies. Within the subject 'Science', introduced in grade 3, children are encouraged to study animals, birds, insects, etc., in their natural environment; the human body, its functioning and its health; changes taking place in the natural environment; etc. Pupils are also helped to understand the usefulness of knowledge in life and are taught to make safe and simple experiments.

In Guyana, the subjects 'Science and health' are taught in all grades. The content is derived from the areas of energy, matter, life, and earth and space, and are interrelated so as to be covered by an integrated approach.

Chile teaches natural sciences in all grades of primary school. The chosen objectives are constructed around an environmental axis throughout primary school. The following themes can be distinguished: scientific processes; living beings and the environment; man as a biological being; matter and energy; and earth and space. The following aspects are stressed: nature, materials, biological change processes, basic needs of living things, and the interdependence between living and other things in the environment.

These examples illustrate the general trend of development. Instead of trying to teach the different natural sciences separately with a view to creating an understanding of particular subjects, the school systems which have accepted an integrated approach have broken down the boundaries between subjects. A set of interdisciplinary themes are selected. These themes vary between countries but are, for example, the environment, the family, space, resources, matter and energy, body and health, school, neighbourhood, seasons, and so on. A common characteristic feature of these themes is that familiar facts can be used in teaching and contribute to the creation of an understanding of these phenomena. The themes are not necessarily confined to dealing with scientific facts, but social studies, language and humanistic ideas may be combined in the presentation. From the study of these themes, the pupil is expected to accumulate essential information, to approach phenomena in a scientific way and to develop a scientific way of thinking and solving problems. In later grades, teaching can become more theoretical and subject specific.

Two examples demonstrate the way science education is covered in countries employing polytechnical education.

In the German Democratic Republic, only the subjects German, mathematics, manual training, gardening, drawing, music and sports are found in the first four grades, while biology and geography appear in grade 5, with the other sciences following. Teaching is thus centred around subjects. There is, however, study of both the social and the physical environment, beginning in grade 2. Science courses are said to lay the groundwork for specialized instruction in the natural sciences starting in grade 5.

A close link with real life in nature and society is the characteristic feature of biology instruction in grades 5 and 6. The pupils are not supposed to adopt a utilitarian approach but to acquire theoretical knowledge in a systematic manner whose relevance to real life is illustrated by practical examples. On the one hand, theoretical knowledge is imparted and the pupil learns how to apply it in production, technology and everyday life. On the other hand, he is faced with practical problems and with finding a way of penetrating and substantiating these problems in a theoretical manner. The subject is divided into sub-courses in grades 5 and 6. Physics instruction starting in grade 5 is designed to extend specialized science education. It is related to mathematics instruction and to other natural science training. The physics branches introduced in grade 6 are extended in grades 7 to 10. The content is structured along technical and educational lines: matter, principles, thermodynamics, optics, etc. First an empirical basis, as broad as possible, is created, leading to the further teaching of abstract concepts.

In the USSR, nature study is included in grades 2 and 3. The pupil learns to find directions by the sun or by compass; read a map or plan; describe certain properties of minerals, water and local soil; keep a nature diary; use laboratory equipment; establish simple relationships between animals, plants, the environment and human labour; maintain personal hygiene; protect plants and animals; and to treat nature with respect.

Content of technology education

A few examples will be given of the way in which different countries state that technology education is taught in their schools.

Denmark mentions the subject 'Crafts', found in grades 4 and 5. It aims to develop manual skills and teach techniques which provide experience with tools. The subject gives opportunities to demonstrate personal qualities different from those learned in theoretical subjects. The pupils use tools, acquire skills and experiences of immediate value, and gain a basis for positive leisure-time activities. They learn to use correct work postures and movements, how to use handbooks and written instructions, and how to make drawings and calculations before starting work. The course includes woodwork and metal work.

In Paraguay, the integrated 'Nature, health and work' subject includes a theme 'Child and work', which teaches technology. It deals with work at home, and with agricultural and other work outside the home, and is meant to contribute to the building up of self-knowledge, creativity and responsibility in the child. When the integrated subject splits up into its separate parts in grades 4, a 'Work' subject appears. It includes the following areas: conservation of soil; cultivation of vegetables, trees, fruits and flowers; preparation of food; administration and improvement of the home; and projects such as the breeding of domestic birds and work in a factory.

In Turkey, both in the existing and in the new curricula, technology is taught within the subject areas 'Drawing and work education' and 'Technology education'. Among the units covered by the former are manual work, work using different materials, graphic work, etc. 'Technology education' is divided into agricultural education (activities in the immediate neighbourhood, nature protection and agriculture), work and technical education (graphic, ceramic and woodwork; electricity, energy; industrial knowledge; etc.), commercial education (commerce, commercial mathematics, typing); and home economics (administration of the home, clothing, nourishment, manual arts, family economy).

Among the eight programmes in primary school in Mexico, one is called 'Technological education'. Here, the pupil learns simple technological processes, is taught to plan and design useful objects, analyses their ways of functioning and develops a participative attitude. Technological education in grades 3 to 6 deals with three themes: energy, communication and production processes. The training is said to be closely related to the science stream. In grade 3, solutions are sought to concrete problems by means of technological processes. Man and animal power is mentioned as a primary form of energy, the great advances of technology are discussed, and ways of replacing muscle power with machines are described. Various forms of communication are presented. The cultivation of vegetables and the breeding of animals are also covered. In grade 4 simple machines are studied, as are tools and materials. Cultivation and the use of fertilizers are dealt with. In grade 5 the pupil develops a capacity for particular skills and is involved in manual work and mechanical construction. The idea of research is discussed, and work plans and team work are brought up. In grade 6, the pupil is taught to plan and construct objects of his own design. Stress is placed on his sense of judgement. In the energy field, emphasis is placed on electricity, light and solar energy, and in the communication field on the use of electricity in various ways.

The German Democratic Republic places heavy emphasis on subjects like manual training, gardening, polytechnical education and drawing, and the significance of their social benefits. Through manual training, pupils are introduced to knowledge of technology, learn about basic forms of work organization, obtain an insight into simple methods and processes, and become familiar with materials, their structure and their macro-economic significance. They are thus provided with the foundations for the polytechnical instruction which follows and have meanwhile acquired essential knowledge useful in other subjects. Through productive work, polytechnical education starting in grade 7 engenders a certain loyalty in the pupils towards their enterprises as well as a sense of responsibility, and it has a positive impact on their work skills. Their understanding of basic relationships in production and of work methods is deepened. School gardening is also considered a polytechnical subject, covering: plant cultivation methods; facts about plant growth; soil, climate and human activity; planning, preparation for and evaluation of work; experimentation; moral conduct during work; and safety and health protection.

In the USSR, 'Labour education' (all grades) teaches the pupils the names and uses of tools and how to make things, develops drawing ability, teaches them to assemble working models by following technical instructions, and provides other training. Agricultural work implies raising seedlings, checking seeds for germination, planting flowers and vegetables, propagating and tending house plants, gathering produce and keeping records.

Technology education differs between countries. In some cases it seems to form an applied or manipulative part of science education, exemplified by the study of simple machines, various communication devices and the study of electricity and other forms of energy. In other cases, heavy stress is placed on learning to use tools and on developing manual skills, proper work habits and an awareness of different materials. In still other cases, the emphasis is placed on learning actual economic skills in agriculture, gardening, industrial workshops and offices. There are also cases where it is found important to teach work organization, team work, co-operative methods, the value of work as such, and the relationship between school, practical work and societal development. In a few countries, not mentioned in the above discussion, advanced skills like computer programming are taught. To a certain extent, these differences are correlated with different ways of teaching technology, either as a separate subject or as a part of different subjects or areas; educational and political philosophies may also have an influence. It is evident, however, that no analysis exists of what technology content ought to be taught in different situations.

The greatest change in primary school curricula during the last few decades could well be the introduction of practical work in pre-vocational education. While this is meant to give children an opportunity to develop attitudes and values for their future life in society, it should also allow ideas from the outside world to influence school systems, thus making the school more involved with everyday problems. An effective link may be created between school and society and between theory and practice. This development has been strongly supported by international organizations and meetings of responsible politicians. Already the Conference of Ministers of Education and Those Responsible for Economic Planning in Arab States held in 1977 [60], drew attention to the relationship between school and work. Other conferences of the same kind, for example those held in Mexico City in 1979 [63], in Sofia in 1980 [7] and in Harare in 1982 [6] emphasized the necessity of creating a close association between school and community life, and between the acquisition of knowledge, skills and attitudes, and productive work. These conferences recommended that children be acquainted, from an early age, with the value of practical work and productive activities.

In different countries this practical or pre-vocational training is known by different names; it may form part of subjects like polytechnical education, technology or life education. Evidently, it overlaps with technology education; in some cases it is hard to see any difference. The proportion of the timetable devoted to practical work varies considerably. In many countries it is non-existent; in others it means simply a superficial introduction to different kinds of work; and in still others it may occupy 5 per cent or more of the timetable. The latter is frequently the case in countries with polytechnical education. Some developing countries seem to show a great deal of interest in this aspect of education, especially in the field of agriculture, while so-called Western countries rarely stress it.

Justifications for the inclusion of practical work and pre-vocational education at the primary level have been expressed by many Member States. The socialist countries feel that education should prepare children for socially useful work and that there should be a direct link between the school and society at large. Other reasons propounded are that it promotes education related to production which is useful to the development of the economy, or even that it is possible to make the schools party self-supporting. In still other cases it is emphasized that the schools shall be integrated into their environments and meet local needs; this means changing pupils' attitudes towards societal phenomena and producing the necessary skilled manpower at the same time.

Further discussion of practical and pre-vocational education and its usefulness may be found in a series of Unesco publications [29, 32 and others] and in the *Yearbook* written by Holmes [22].

CONCLUSIONS ABOUT THE EXISTING SITUATION

A few trends in science and technology education ought to be noted.

In most countries, little time is devoted in the school timetable to science education and still less to technology education. Possible reasons for this have been suggested, although the most valid answer seems to be the lack of stress on these subjects in the statements of objectives discussed in Chapter III. Technology education as a separate subject is a new phenomenon. In spite of efforts by organizations like Unesco and in spite of its obvious practical importance, it is often disregarded by educators and politicians. The situation is particularly bad in areas where the environment does not offer opportunities for incidental learning at home or in workshops, and where little information is provided by the mass media.

Science is still taught mainly in the upper grades of primary school, but there are some trends for science to be taught in the lower grades as well. This is coupled to another phenomenon which concerns primary education: integrated subjects or curricula. In the lower grades, major interdisciplinary themes with relevance to the pupils' local situation allow the teaching of scientific skills, facts and ideas which can later be developed into a theoretical structure within each discipline. This form of teaching can be found in all parts of the world, although with different names. The intention is to allow the pupils to see education as a means of finding solutions to important problems, which in itself means that the school becomes a relevant institution. The socialist states have chosen polytechnical education, a slightly different solution, but the stress on socially relevant knowledge and skills, and on co-operation between school and society, school and work, and theory and practice plays a similar role.

Technological education does not yet seem to have found its final shape and specific role. It is found within different subject areas, it is designed to meet different needs, and it has different content in different countries and even in different schools within the same country. To some extent the picture may be distorted by the fact that the word 'technology' and the corresponding words in other languages (*les techniques, la tecnología,* etc.) have not yet been given a fixed meaning in educational contexts. (See also *New trends in primary education,* volume I [19].) It is essential to establish what needs shall be met by the subject, what goals and objectives shall be reached by this education, and what educational content the curriculum shall contain in a country, as well as to establish its relationship with other subjects.

The question of what and how to teach in science and technology courses in primary school has been discussed by educators. Major research and development projects, in developed as well as developing countries, have dealt with this question [19]. As is always the case with the results of such efforts, it is hard to determine the immediate effects in ordinary schools, where there is often an evident lack of facilities, equipment and well-trained teachers. No new and major results of the on-going discussion and the studies made seem to have happened, and it is difficult to perceive the general rules or laws that may yet appear. The discussion itself is important, however.

Black [3] stated a few propositions in the selection of a science teaching policy, some of which may be relevant to the present discussion: 'We need to start from careful study of children's present attainment'; 'We need to clarify the relative importance and interactions of process and content'; 'We need a comprehensive model of the curricula in order to locate and interrelate work on science, on technology and on other interdisciplinary areas.'

The first proposition refers to the background and abilities of children. Substantial studies have been carried out by educators and psychologists about the intellectual and other development of the child. Suffice it here to refer to the Piagetian concept, which has determined that the child passes through definite stages of intellectual development. Children in primary school are, according to this theory, at a developmental stage during which they are not able to perform logical reasoning without reference to concrete materials. This influences the way in which science materials can be presented and places heavy stress on 'hands-on' learning and laboratory work. In the information provided by Member States there is repeated reference to the need for a curriculum which is adapted to the child's level of development. Black refers to studies carried out on the extent to which children are, in fact, able to carry out the tasks asked of them. He also points to a need for some model of the learner and the learning process. On the other hand, studies quoted by Werdelin [88] show that the way in which problems are presented and the manner in which theoretical reasoning is anchored in concrete facts is of great importance to the concept formation taking place.

The question of whether to stress content or process in the teaching of science has also been widely discussed. Traditionally, of course, teaching in primary school placed heavy emphasis on knowledge of facts, often learned by heart without reference to a whole. This means that the curriculum was extremely content oriented. Many modern educators would be more inclined to state that a main aim of education is to teach children to find out for themselves and to make them approach problems in a scientific way; the curriculum should be process oriented. The elements of the process approach are well known: classifying, measuring, inferring, etc. They have been stated among the objectives of science and technology education by many Member States, as summarized in the previous chapter of this book.

It is hard to conceive of a curriculum which is entirely process oriented. Technology education must be essentially content oriented since it usually aims to teach particular skills, the properties of certain materials, etc., without entering into explanations. In science, there is knowledge which is essential to our very survival, for example, knowledge of our body and its functioning, conservation of nature, different forms of energy, dangers brought about by the use of science and technology, and so on. Other such knowledge belongs in the general cultural backround we should all acquire: means of measuring length, volume, weight, time, temperature, etc.; the use of simple machines; communication media; a scientific vocabulary and an understanding of corresponding concepts; and so forth. On the other hand, it should be realized that it is possible to teach only a few facts and that it may be preferable to provide the child with the means of acquiring new facts. Process-oriented curricula may form such a means. The question of content and process is discussed by authors like Harlen [20], Young [94] and Black [3]. Many countries have shown a clear tendency towards process teaching. Terms like 'understanding', 'scientific methods' and 'scientific thinking' appear frequently in their reports.

In the above text, the science education model developed in Colombia demonstrates how different disciplines can become part of an integrated curriculum. Other models may exist, and in other countries there might be interdisciplinary themes or projects without an overt model. In most cases these themes, as well as the materials used in the teaching, are taken from the surrounding environment, and are based on the experiences and the problems encountered by the pupils. There may be a difficulty, however. Problems in the environment are not necessarily simple, and a simplification may make the teaching abstract and even meaningless. The environment and the pupils' sphere of experience may also have severe limitations and it is essential that education goes beyond the limited world of the child and prepares for roles in a larger society. This means that education in the schools must, sooner or later, be based on other materials and try to solve other problems that those near at hand.

CHAPTER V
The delivery system in primary school

THE PLACE OF THE DELIVERY SYSTEM IN THE EDUCATIONAL PROCESS

Educational goals and objectives, as discussed in Chapter III, are expressions of aims behind the educational process as accepted by politically responsible authorities and influential individuals. The goals and objectives are further concretized in two ways: by statements on curriculum structure and content; and by the creation of a delivery system. The former was discussed in Chapter IV, the latter will be discussed in the present one.

Several separate parts or elements can be distinguished in the educational delivery system: teaching methods, educational organization and administration, examination and guidance systems, teacher training systems, and so on. In addition, the co-ordination between curriculum structure and content and the elements of the delivery system, and the co-ordination among the latter elements may greatly affect educational outcomes.

Like curriculum structure and content, the delivery system is dependent on educational goals and objectives. Elements of the delivery system are sometimes even mentioned among the objectives, such as when an objective is to use group work or an individualization method, to employ more highly qualified teachers, or to abolish examinations entirely. (See also Chapter II on tertiary goals.) The delivery system should be directly dependent on the objectives, since its criteria ought to be determined on the basis of the latter, but in reality this is far from always the case. More often, the relationship is indirect: the delivery system is such that educational goals and objectives are reached as far as the existing situation permits.

A model showing the relationship between the main groups of factors which determine educational outcomes is depicted in Figure 5.

In this model, there is one concept which needs interpretation: 'Frames'. It stands for restrictions, constraints and difficulties encountered in any situation, for example: economic, material and other often very limited resources; lack of skilled and adequately trained teachers and others staff; attitudes in different groups; and agreements, laws and administrative considerations.

To identify 'frames' with 'difficulties' and 'restrictions' is not quite correct either. The frame around a painting not only delimits but also indicates that a certain area is, indeed, available. An economic frame not only indicates that resources are limited, but also tells us that certain resources do exist and can be used. The frames are therefore indicators of the situation in which the system works: what material and human resources exist; what people think and feel; what laws and agreements are found, and so on. Difficulties are often implied, of course. To give an illustration: the ability of a school system to improve its science teaching would depend on material resources

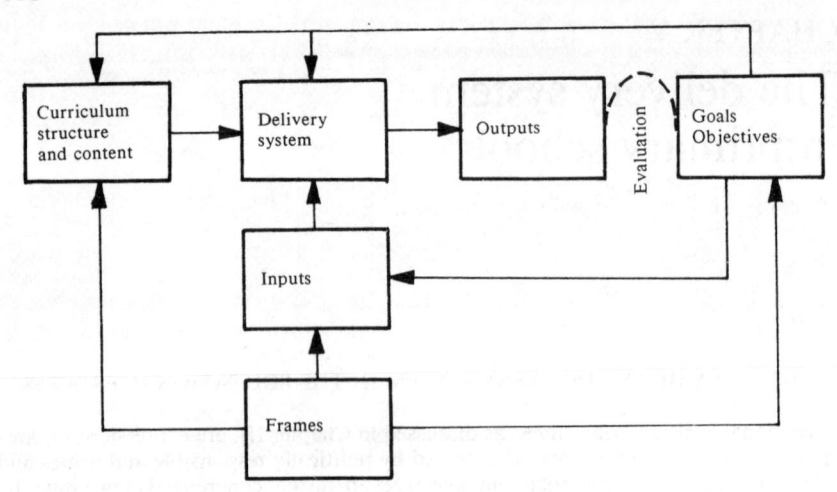

FIGURE 5. Main factors determining educational outcomes

available to build laboratories, buy equipment and books, etc.; on the skills and experience of the teachers who are responsible for the teaching; on the attitudes of students, parents, teachers, administrators and others towards science and towards different teaching methods; and on regulations governing the situation and agreements between the school system and the teachers' professional organizations.

The structure and content of the curriculum are highly dependent not only on the goals and objectives determined but also to a fairly large extent on frames like attitudes, laws and agreements, the availability of teachers and material resources. The delivery system is also dependent on goals and objectives, although often indirectly via the stated structure and content of the curriculum. It is highly dependent on frames of all kinds. The way a school subject is taught is to a very large extent a result of material investment, teacher qualifications and experience, and so on. This relationship is essentially indirect: the delivery system does not depend on resources available but on resources spent, not on teachers available but on teachers employed, and so on. This is indicated by the box 'Inputs' in Figure 5.

The effects of the delivery system in an educational situation are measured in the form of outputs: student performance, attitudes, skills, etc.; teachers' work loads, attitudes, etc.; changes in attitudes and the values of different groups; and so on. The relationship between the stated goals and objectives, expressed as criteria or expected outcomes, and the real outcomes is measured by an evaluation process.

According to this model, the delivery system occupies a very central position. On the one hand, it exerts direct influence on the outputs; on the other, it can be manipulated by the system to a much higher degree than the structure and content of the curriculum. In this chapter, an attempt will be made to present an overview of the delivery systems in different parts of the world, particularly in respect of science and technology education. Lack of reliable data as well as limited space place certain limitations on the discussion which will concentrate on those essential areas where data are available: teaching methods, administration and class organization, evaluation of performance and teacher training.

A great deal of research has been devoted to studying the importance of factors in the delivery system. It is not feasible to present a critical review of research findings in this connection, but reference to handbooks and similar publications will be made in the following text, and a few important research results will be quoted.

TEACHING METHODS

Superficially at least, the methods used to teach science and other subjects in primary school have changed a great deal during recent decades. While pupils previously sat in neat rows listening to a teacher who explained and demonstrated scientific phenomena and facts, they are now often organized into groups that try to solve problems themselves. New teaching materials have been produced so that the pupils themselves are able to investigate phenomena, and laboratory work often plays an essential part in the learning situation. In most countries, audio-visual aids, ranging in complexity from a blackboard and a few posters to a computer and a video recorder, are assumed to improve and facilitate the teacher's task. Practical experience in workshops, factories and farms, and demonstrations in the field during excursions and study visits further strengthen the learning situation. Technology education should stress the pupils' own problem-solving activity and practical skills even more than science education.

But has the teaching situation really changed very much? If changes in the classrooms simply mean a new seating plan while the teacher continues to explain and demonstrate, if the new teaching material turns out to be too expensive for regular schools, if audio-visual equipment gathers dust on shelves or its use becomes a haphazard affair, and if the excursions and the practical experiences are badly planned and turn into jaunts, very little real change has taken place.

The area of teaching methods has been a favourite one for educational researchers and, after the teaching of languages and mathematics, science teaching has drawn the greatest attention. Literally thousands of books and scientific articles on the topic have been published; to take a single example, the publication *New trends in primary school science education* [19] refers to 255 recent works. A great deal of knowledge exists and should influence teaching all over the world.

Studies on the effects of teaching methods have been summarized in handbooks of educational research like the ones issued by Gage [15] and Anderson [2]. Research on ways of presenting materials to be learned was made the subject of a critical survey by Werdelin [88]. Among findings of interest to the present discussion, the discovery method of learning in the classroom can be mentioned as a viable alternative to the usual teacher-directed learning process, if satisfactory learning materials are available. It is also known that teaching materials and instructional aids not only facilitate but sometimes even make learning possible. This is particularly the case when abstract concepts are being learned in primary school: the pupils are at the concrete-operational stage, according to Piaget, and concepts to be learned must be anchored in concrete experiences.

Due to the nature of the available material, it is not possible to make an in-depth study of the teaching of science and technology in all Member States. What can be done is to investigate the extent to which certain innovations have reached different countries.

Classroom procedures

In 1980, experts on the incorporation of science and technology in the primary school curriculum [47] made concrete suggestions on how to teach science. In each general topic and at each grade level, a science cycle of four stages was suggested. This is shown in Figure 6.

The purpose of this model is to develop the habit of thinking on the part of the pupil at each phase of the cycle. As is shown in the figure, the full cycle has four structural features which aim to develop analytical, evaluating and planning skills. These features are: finding the way of knowing facts, applying these facts, seeking consequences and studying their value. In each case, questions must be asked. In order to find ways of knowing and applying facts, the pupil must make observations, investigate the phenomena, interpret the information critically, and ascertain whether new studies are needed. The scientific work of description, explanation and control becomes essential. It is assumed that instruction that follows this cycle is likely to produce young people who can participate in societies which require an increasingly sophisticated understanding of science and technology.

In spite of the fact that much of the educational debate concerns classroom behaviour and that Member States were invited to describe methods of teaching, not all countries have described innovations in this area or even given an account of the existing procedures. However, a sufficient number of countries have sent in reports which allow some statements to be made about developments in teaching methods. There is a great deal of agreement between the reports. Many countries emphasize that a main aim of science and technology teaching is no longer to provide knowledge, which may soon become outdated, but to teach understanding and skills; this is evidently another aspect of the shift from content-oriented to process-oriented curricula. For this reason, the pupil is taught to observe, measure, classify, conduct experiments and take other steps in the scientific process. 'Hands-on' methods, practical experiences in workshops and on farms, etc., are emphasized; an 'activity pedagogy' is introduced. This can take different forms, of course, depending on national and local conditions.

In Peru, the following considerations are observed when determining the teaching method: adaptation to the maturity of the pupil; use of physical activities; participation in self-learning and group learning; and use of environmental resources. In science and technology education, stress is placed on the intellectual development of the pupil. He is given opportunities to observe, classify, compare, suggest explanations, experiment,

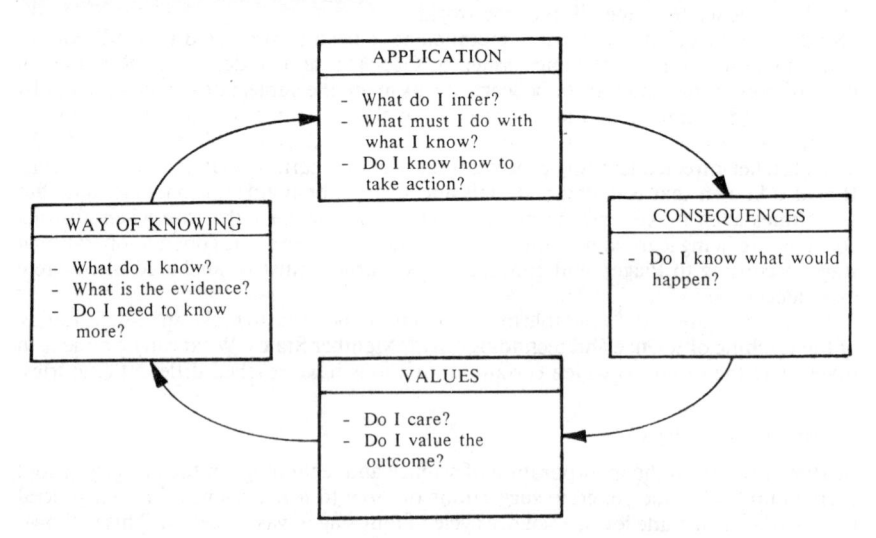

FIGURE 6. The science cycle, as depicted in a Unesco publication [47]

The delivery system

draw conclusions, generalize and communicate. Motor and behavioural developments are also emphasized.

In Paraguay, the introduction of science and technology is an activity-oriented process. The pupil observes, makes experiments, does group and practical work, and draws conclusions after having gathered enough data.

The methodology suggested by the new programme in Chile places the stress on the learning of facts and the development of skills, abilities and attitudes through the study of simple objects and systems. The pupil observes, measures, classifies, draws inferences, makes predictions and so on, and communicates his/her discoveries. The content of study should not be limited to the description of phenomena, but should imply the necessity to observe, measure, etc.

In Colombia, the curriculum renewal effort presently going on concerns all features of the curriculum, including methodology. It should permit an adaptation to the social and cultural circumstances of different groups and to the developmental level of the child. Stress is placed on skills and ability, not on memorization. Active, permanent and critical participation of the pupil is desired, and the on-going participation of the parents and the community is welcomed.

In Jamaica, activity-oriented science and technology curricula, relevant to and based on the local environment, are written. The explicit policy with regard to science and technology education at the primary level is to introduce children in a 'hands-on', meaningful and practical way to the process of the disciplines.

An answer from Pakistan mentions the orientation towards a new science and technology methodology, and the attempts being made to introduce a guided-discovery approach in primary school. It is stated, however, that the renewal of science and technology education has only just begun.

In Gabon, there is also a renewal of the educational process adopting a very broad approach. The observation method is used within a multi-disciplinary field. All senses are employed to awaken the scientific spirit in the pupils, who should spontaneously raises questions about their environment. Such a method leads to curiosity, questioning, the need to know, hypothesis building, etc. The pupil then verifies hypotheses, either systematically or by going back to past experiments, organizing data and taking other steps in the scientific process.

Uganda reports that its efforts to ensure an appropriate introduction to science and technology education take place through teacher training. The curriculum renewal concerns itself with the methodology used. The teaching approach in the schools is child-centred and geared towards problem solving, as well as the practical application of knowledge and skills to environmental challenges.

The development of teaching/learning methods in Nigerian schools has de-emphasized the memorization and recital of facts and encouraged practices like exploration and experimentation. Training for manual skills is also stressed.

In Mauritius, the basis for science and technology education is related to the level of scientific and technological development in the country and to the children's backgrounds and level of development. The aim is to provide tools of inquiry rather than areas of knowledge. Education in school shall provide a vital link with educational experiences at home, in the community and in work situations. Activity-based methodology is used.

In Jordan, there is an effort to make maximum use of the local environment; the use of activities is emphasized and training is made pertinent to the local situation.

Iraq aims for a general renewal of the primary school curriculum. Workshops and school farms are developed to involve children in work, the pupils' creative abilities and skills are trained enabling them to take part in practical and artistic activities, while extra-curricular skill-related activities are encouraged.

In Australia, each state sets its own goals and objectives, and determines its own curricula. In South Australia, for example, the objectives of science and technology education include the development of knowledge and understanding of chosen content and the creation of attitudes of scientific inquiry and skills, as well as attitudes towards good and responsible uses of technology. Science courses taught in school vary considerably. In many cases, they are integrated with courses within other areas. Accordingly, the methodology used also varies, but a widely accepted example is the one in the centralized Queensland curriculum which emphasizes five key elements matched to children's development at primary level: success skills; manipulative skills; positive attitudes; concepts; and 'hands-on' activities.

The United Kingdom draws attention to the necessity of carrying out research in school rather than just hearing about it. The use of scientific investigations, the design of experiments and instruments, and the formulation of hypotheses give a particular character to some of the work.

There is no national curriculum in the United States either, but many different strategies are being employed. At best, it is possible to distinguish trends. The following quotation from the response to the questionnaire on science and technology in primary education may illustrate one:

In recent years, there has been substantial development in our understanding of classroom and school practices that can improve student achievement. This 'effective schools' knowledge has been gained largely from studies of elementary reading and mathematics, although broader implications can be drawn, and it is likely to influence primary education in a number of ways over the coming years. Changes can be forecast chiefly in three areas: (1) effective classroom teaching practices, (2) coordination and management of the instructional program at the building level, and (3) the shared values and culture at the school.

It may be of interest to compare these data with those obtained by the study of technology education in thirty-seven countries [84]. Here it is stated that few countries mention methodological facts. The 'learning-by-doing' method was obviously the most popular, at least at the primary level which concerns us here. Projects in workshops are the main focus of interest. In most countries, practical abilities within technology education courses are simply utilitarian, based on the execution of projects; but in a few countries with a high level of industrial development, exercises are given in accordance with a method that fosters the gradual formation of concepts and the grasp of logical thinking.

In conclusion, it can be stated that many countries undoubtedly try to create a science teaching situation which agrees with the one suggested by the expert group mentioned above. Stress is being placed on an understanding of relationships instead of rote learning, and the scientific process and scientific thinking are focal points in the educational process. Key words like 'activity method', 'practical application', 'environmental protection' and 'skill training' also appear in almost every country report and provide information about the type of science and technology training that countries want. This does not mean that all teachers in all countries always use this modern methodology. On the contrary, there are reasons to believe that the traditional teacher-centred delivery system is still common and that knowledge is a main aim. Several countries have stated that methodological renewal is rare or slow.

Instructional resources

Science means 'doing something with something', even if only observing it, but also learning something in the process. Technology means 'changing something into something else'. From these truisms can be drawn the conclusion that science and technology education means that pupils should have objects to handle and work with. Language and arithmetic deal with abstract concepts which must be learned and kept in mind; many other subjects deal with facts which can be memorized or stored in books and elsewhere. Science and technology deal with concrete objects that can be observed, and a sample of these objects should be available for study.

Like other disciplines, science and technology are characterized by a mass of information which must be learned or kept available in written form. The stress on understanding rather than memorization in science has meant that the problem of storing and retrieving information has been accentuated, and a great deal of development has taken place in science as well as in other disciplines concerning ways of keeping information available. Books, simple audio-visual aids and complicated technologies are essential in the teaching/learning situation. For obvious reasons, much of technology education as it is at present conducted in most countries becomes fairly independent of such instructional resources, but at a more advanced level of education and when advanced technologies are studied, these resources are essential.

Instructional resources of two kinds thus appear in science and technology education: examples from the world studied; and means for storing and retrieving data. The two sets of materials have an intersection, however: many of them are both objects of study and means for storing and retrieving data. This is the case with films, overhead and other projectors, radio and television systems, etc., as well as with advanced technologies like video systems, computers and communication satellites. In the following text, the instructional resources will be sorted according to their characteristics rather than according to their functions. Some of the data presented have been taken from information collected by means of the International Conference on Education's study of trends in education [35].

Books, pamphlets, papers and other reading materials form basic instructional resources and provide the most commonly used means for storing data. A number of countries have mentioned a need for improving such materials.

In Ethiopia, the new curriculum is still at the experimental stage, but means are sought to raise educational standards. Among the steps taken is an improvement in the quality of teaching/learning materials. All materials, including textbooks and teacher guides, are tried out in experimental schools, including new texts on environmental studies and science. In Iraq, the educational renewal has meant that textbooks have been revised and teacher guides produced. The Syrian Arab Republic reports considerable work to produce textbooks. Committees have been appointed to issue them, in-service training has been organized to introduce them, and they have been revised after being tried out in schools. In Bulgaria, the new educational programme with integrated science and technology is supported by means of textbooks, teacher guides and other materials. The renewal of programmes and textbooks in Poland started in 1978/79 in the lowest grades. By 1985/86, the primary school programme and the textbooks will have been entirely modified. China has re-edited reading materials in nature study courses as a result of experiments. In Pakistan, hundreds of different textbooks have been revised and approved by the Ministry of Education.

The need for reading materials other than textbooks is pointed out by a number of countries. Guinea and Kenya stress the need for the publication of newspapers in local languages to maintain literacy, and the United Republic of Tanzania points to the need for reading materials supporting, for example, agriculture. Jamaica states that the use of radio and television in school has increased the demand for reading materials. Nigeria mentions school libraries and states that in-service training is given to teachers in the management and organization of such libraries. Kenya has established libraries in all major towns.

It goes without saying, however, that most countries feel a need for textbooks and other reading materials to improve the literacy skills of their peoples; science and technology are rarely mentioned. Some countries state that textbooks are provided only for mother tongue and mathematics. The scarcity of textbooks and other reading materials is often appalling. For example, the experimental school system in Colombia, *La escuela nueva*, has produced significant textbooks, but it is anticipated that the country can only afford five sets per class.

While reading materials like books are used in science and technology for storing and retrieving information, most simple audio-visual aids are used to demonstrate scientific phenomena and allow pupils to manipulate the material. The group of experts on the incorporation of science and technology in the primary school curriculum [47] discussed this type of material at some length. They suggested that teachers should obtain from their locality many of the materials which they might need. They could even prepare and manufacture their own sets, particularly in the early grades. The experts stated, however, that many teachers are not aware of the possibilities of 'junk' — this is the term they used — and not trained to use it. They also found that science kits are essential as they can provide practical experience. But kits have drawbacks: they may be costly, and few teachers are trained in their use. The use of objects that can be found in or are produced by the locality is mentioned as an additional source of audio-visual materials.

Few countries have provided data on or otherwise mentioned simple audio-visual aids. Nigeria states that her educational renewal involves the development of such aids for primary schools. There is a National Educational Technology Centre to promote the introduction of science and technology at the primary level, and an Educational Resource Centre Network has been established all over the country to provide schools with technical services.

Science kits and equipment have been developed. Pakistan reports that teaching kits for primary schools have been developed and provided to 50,000 primary schools. In China, the catalogue on equipment available for mathematics and teaching about nature has been revised by the Ministry of Education. It is found that the quality of the equipment has improved, while local authorities are required to provide what is still lacking. In Hungary, one step to increase the efficiency of science and technology education has been to purchase educational materials and experimental appliances. The Syrian Arab Republic states that science and technology programmes have been enriched by technological equipment which allows the pupil to conduct experiments. Among steps adopted in Iraq is the development of workshops and laboratories in primary schools and the encouragement of school farms in rural areas. Madagascar has created a centre charged with equipping all schools, principally primary schools, with suitable teaching materials.

The Unesco study of technology education as part of general education [84] gives data on equipment and materials used in this subject. Only a minority of countries have special rooms for handicrafts in primary school. Workshops are sometimes found in special centres. There is often a lack of equipment and material resources.

The more advanced technologies used in education include radio and television, video discs, computers of different kinds and satellite communication systems. All of these are both instructional resources, scientific phenomena and objects of study. They are rarely included in primary school courses, however. In the study of educational trends, quite a few countries reported efforts in this field.

The most commonly used of these technologies are radio and television, primarily the former. They have two main uses: to support teaching in the classroom; and to facilitate distance education. In developing countries, great expectations are placed on literacy training and other forms of general education based on these media.

The delivery system

In African states like Ethiopia, Guinea, Kenya, Nigeria and the United Republic of Tanzania, radio broadcasting is used in the primary school or to promote and support literacy. Ethiopia states that more than 5,000 schools receive radio programmes. In Kenya, radio programmes cover general knowledge, health and nutrition, the environment, etc. Over 50 per cent of all programmes of 'Voice of Kenya' radio are of educational value.

In South and Central America, there is a wide range of uses for both radio and television. In Argentina, 'La Telescuela Técnica' has transmitted both general programmes and programmes on science and technology in the country and on industrial technology. Since 1977, an officially subsidized programme for the improvement of science teaching in primary school has been operating. Distance education at an advanced level exists. Chile has created packages for basic education, and also sends programmes to provide vocational education using self-instructional textbooks. Colombia provides literacy training, primary education for adults and more advanced training via radio and television networks, while Cuba uses the latter for training in different subjects and for the teaching of communication skills and proper behaviour. In Mexico, the main aim of radio and television courses is adult education. Secondary education courses are also given, however, particularly for remote areas where secondary schools have not yet been built. Jordan and Mauritius find that radio and television promote self-education and, accordingly, they accord them high priority.

Poland states that the new educational plan takes into account co-operation with radio and television networks. This results in programmes designed exclusively for primary school pupils. Hungary has also established close co-operation with these media. In the USSR, special centres prepare television programmes for schools. They are only conceived as aids, however, used to widen the child's sphere of experience and promote positive attitudes. In the German Democratic Republic, radio and television are widely used for educational purposes. They are intended as instructional resources and do not replace the teacher.

There are very few countries which do not use at least the radio network for educational purposes, and even regular programmes for the general public provide children with a fund of information. The Federal Republic of Germany finds that it is possible to use children's experiences in the teaching of some subjects, but it must be taken into account that they already get much of their knowledge and experience from the mass media. Certain problems have been encountered. Radio and television education provide a passive educational situation unless followed up by the teacher. Comparatively little radio and television time can be devoted to each school course. In many parts of the world there is no electricity, or the power supply is irregular, as reported by, for example, Nigeria, Sri Lanka and Zambia. The latter country also points out that few people have an organized system of listening.

Modern audio-visual aids are also reported. In the United States, videotaped courses are available, particularly at an advanced educational level. As in most developed countries, microcomputers are used in the teaching of computer science. Japan reports that video discs have been introduced experimentally giving access to data files. It is believed that they, linked to microcomputers, will become powerful tools for learning. Finland sees the use of computers as an important means towards the individualization of education and in remedial and special education. The Republic of Korea states that the use of microcomputers for educational purposes is being tried out. Ireland has used computer-assisted learning in remedial education. In Denmark, this new technology is being developed for special education, which may lead to quite new approaches in primary education for pupils without handicaps. In Jamaica, computer-based instruction is being studied by ministerial committees and introduced to teachers and teacher trainees.

Computer technology and use is also a field of knowledge introduced in primary school in a number of countries, for example, Australia, Cyprus, Finland, Japan, Sweden and the United States. In Japan, only a few primary schools have their own computers, but they are very common in upper-secondary schools. Cyprus has introduced microcomputers experimentally. Sweden has taught computer technology and use for more than ten years on an exprimental basis.

Even still more modern audio-visual aids are reported by a few countries. The United States mentions the interactive video disc, communication satellites, microwave radio systems which allow two-way communication, video-texts, interactive television and educational telephone systems which link several localities. Japan has videotexts for classroom teaching on a wide scale, and Sweden reports text telephones for the teaching of deaf people. Australia has a loan video system for education. Some countries, besides the United States, mention the use of satellite communication for educational purposes, for example Australia, India and the USSR. In most cases its use is delayed for political reasons, however: such satellites allow transmission to several countries at the same time and may lead to the undesirable spread of information.

Teaching methods based on the use of the new media and teaching that deals with the media themselves are both expanding fast, partly due to the fact that hardware items, like computers, and the accompanying software are becoming fairly cheap. The use of advanced technologies for educational purposes is generally found to have advantages: they allow the teacher to do his work better, they extend education to remote areas, they may raise the quality of education by calling for better textbooks and better teacher preparation, and they are fairly cheap if used on a large scale. On the other hand, they have considerable disadvantages: the initial investment called for is high; many people resist their use since it is not the 'right education'; large-scale dissemination may endanger local and even national cultural features; and distance teaching does not always provide efficient results. It is generally agreed that the teacher cannot be replaced, and the application of advanced technology calls for a very high level of training on his part to be of any use.

The introduction of advanced technology as an object for teaching has also met sometimes with resistance; quite naturally it confronts the same attitudes as advanced technology in society. There is some danger to science and technology education if the training in, for example, computer technology becomes too dominant in primary school: the school system and the individual school may be inclined to reduce the teaching of more elementary science areas and try to skip a stage in technology training by turning directly to advanced materials. Even in an advanced society, pupils need to learn to understand basic facts and procedures, and to use tools and do simple technological work. On the other hand, computer technology will soon become a necessity for almost any young person who wants a job in an advanced society.

Out-of-school activities

Education takes place outside schools as well: in homes; in playgrounds; during visits to workshops, factories, museums, etc.; and in other places where children acquire experience. It has long been recognized that the school system should make use of such experiences and try to co-ordinate them with those gained in school. In a response to the International Bureau of Education's questionnaire on an appropriate introduction to science and technology, the United Kingdom pointed out that in a highly developed country there is a wide range of commercial and domestic activities which encourage science and technology training.

The group of experts on the incorporation of science and technology education in primary education [47] also stated that in many countries there is a rich variety of out-of-school activities related to science education: field or environment study centres for children; school farms; science fairs, exhibitions and clubs; and studies of the night sky. They also stress the teacher's activities as a co-ordinator, stimulator and adviser, and the fact that specialists like agricultural extension agents, directors of study centres or scientific advisers may be involved. Finally, they point to the importance of the mass media and other attempts to popularize science. The teacher must capitalize on knowledge acquired in this way.

The development of practical work as an essential part of the curriculum, discussed in Chapter IV, offers another means for strengthening science and technology through out-of-school activities. This has had consequences for the way in which the school system operates. As was mentioned in Chapter IV, many countries report that such work forms an essential means for technology training. A document for the thirty-eighth session of the International Conference on Education [29] relates various arrangements for liaison between education and the world of work: management committees; representatives from working life as consultants in governmental education committees; consultation with such representatives; consultation with workers' organizations; participation of local enterprises in the management of schools; local education committees created specifically to promote co-operation between education and working life, and so on. In these circumstances, it is to be expected that some kind of co-ordination takes place between teaching in school and training given in enterprises; in a country with polytechnical education this is a cornerstone of the educational edifice. Science and technology should profit from such co-ordination.

Information about out-of-school activities of importance to science and technology was collected from Member States.

Many such activities can be brought under the heading of *exhibitions and excursions.* The importance of science fairs and exhibitions, whether national, regional or internal to the school, where the pupil can see the results of scientific activities and may occasionally be able to demonstrate his own, was mentioned by, for example, Argentina, Bahamas, Brazil, Bulgaria, Chile, Colombia, Jordan, Poland, Republic of Korea, Thailand and Zambia. The last-mentioned country even reported inter-school competitions. Visits to farms, orchards and other agricultural exploitations were referred to by Bahamas, Bulgaria, Cyprus, Nicaragua and others. Bahamas combined this with visits to industries, exhibitions, telecommunication buildings, naval ships and offices where computers are used, that is, as part of a systematic tour of different places of work which can contribute to science and technology education; while Nicaragua mentioned practical experience in agriculture. Visits to industries, factories and other production centres were referred to by Australia, Bahamas, Cyprus, the Federal Republic of Germany, Gabon, Hungary, Mexico, Paraguay, Peru and Seychelles. Out-of-school excursions were mentioned by Finland, excursions to the neighbourhood and visits to commercial and industrial establishments by Spain, and educational tours by Nepal. Museum visits were reported by Australia, Mexico, the Netherlands, Switzerland and the United States, while the Netherlands reported that special school museums exist in the country. Paraguay made reference to visits to other cities.

Participation in *work-related activities* was reported by a large number of countries. Arts and crafts are often considered related to technology, and some of the out-of-school activities mentioned are closely related to the arts and crafts subject in school. The German Democratic Republic drew attention to handicrafts and construction; Bangladesh to arts and crafts with an agricultural bias; Nigeria to arts and crafts as well as to clay modelling; and the Central African Republic to welding, pottery, woodwork and other crafts. Manual or pre-vocational activities were reported by, for example, Algeria, Cameroon, the Central African Republic, Ethiopia, Jordan, Luxembourg, Mauritius and Peru. Algeria and other countries stated that science and technology training in school is supplemented by extra-curricular activities or participation and/or voluntary work involving productive and socially useful activities. Workshops are said to exist in Chile and Mexico. Agricultural work or other work on farms was reported by countries like Bangladesh, the Central African Republic, Madagascar, Nicaragua and Seychelles; in the last-mentioned country it concerned a school agriculture project. Gardening was reported by Bulgaria; tree plantation by Viet Nam; cooking by the United Kingdom; and work in the home by Madagascar.

A few countries referred to *nature study*. Colombia, the German Democratic Republic, Mauritius and Nepal found it important that children should observe the environment, and the Netherlands organized leisure-time involvement with the natural environment. Poland and Bulgaria mentioned work for the protection of nature; the United Kingdom reported nature trails and field work during holidays; and the Netherlands described trips to biology centres where lectures are given.

Games and other leisure-time activities were reported to be important to science and technology education by a series of countries. The United Kingdom mentioned computer games; the United States referred to minicomputers and stated that their use in the home is beginning to have a marked impact on the learning of the elementary school child. Ireland pointed to home video and microcomputers; the Republic of Korea to computer training courses. More commonly used media like television, radio, the press and films were also mentioned by a number of countries, for example, Argentina, Australia, Bulgaria, Cameroon, Colombia, Ireland, the Netherlands, Nigeria, Paraguay, Peru and the United Kingdom. In most cases, television was considered as very important. The German Democratic Republic mentioned particular journals, books and radio and television programmes with scientific and technological content. Among other answers can be reported those from Nigeria — toys and play instruments; Luxembourg — photography, enamel work, weaving, etc.; and Argentina — photography and computers.

The last cluster of out-of-school activities to be mentioned here are those which can be labelled *group activities*. Some countries find that they may provide knowledge and skills applicable to science and technology. Camps are reported by Cyprus and Finland; scouting by Cameroon and the United States. Youth programmes are mentioned by several countries. The socialist states in Eastern Europe stress their pioneer organizations. Iraq reports that there is a scientific association of 'Vanguards' which trains children in various scientific and technological activities, and Uganda refers to youth organizations like young farmers, home improvement clubs, science clubs, etc. Science clubs are also reported by Argentina, Chile, Czechoslovakia, Ethiopia, Jordan, Seychelles, Switzerland, Thailand, the USSR and other countries. Study circles are found in Nigeria, Cuba, Hungary, Poland and other countries. Cuba states that their objectives are to develop practical abilities and initiate development towards a professional and vocational choice.

For obvious reasons, this summary of out-of-school activities affecting science and technology education does not give a comprehensive picture of the real situation. It can only show that practically all countries organize such activities, without indicating either their frequency, their organization or their effects. It can also be assumed that activities which are not mentioned, such as encounters with science and technology in the home, incidental instruction by friends and adults and via different mass media, or learning by watching others at play or at work, often make an essential contribution, particularly in developed countries.

EDUCATIONAL ADMINISTRATION AND EDUCATIONAL ORGANIZATION

To give a true picture of the system of educational administration in a single country is a considerable undertaking calling for extensive research concerning the institutions involved and their relationships. It follows that such an undertaking concerning several countries is not possible in this context; the present authors must confine themselves to identifying some common trends in the majority of Member States and detecting points where they differ. Even this task meets with difficulties since the available material from countries is very uneven.

A study of school organization offers less difficulty since data are usually available and definite trends can be discerned. A problem, however, is that there are so many aspects of organization that a choice must be made. In this book we will therefore confine ourselves to the organization of classes, since it is felt that this influences the way in which the schools execute their teaching duties.

Educational administration

Certain school systems are said to be centralized, that is, much power is vested in central authorities like a ministry of education, a board of education or some other body, while others are classified as decentralized, which means that local authorities or even the schools themselves have taken over a large part of the responsibility. This dichotomy is not a true one, for even in the former case a great deal of authority is usually delegated to regional and local authorities, and, on the contrary, no system of educational administration is completely decentralized. There are, however, a number of countries which are rather highly centralized while a number of others can be considered typical examples of decentralized ones.

In all countries there is a hierarchy of administrative bodies, from the central government via the ministry of education, through a series of regional and sub-regional offices down to the local board and finally to the individual teacher. The names differ, but the functions of these authorities do not vary very much. A typical description of such a system is provided by the United Republic of Tanzania:

At the national level there is a Ministry of National Education, and within the Ministry there is a Directorate of Primary Education headed by a Director.... Regional Education Offices are responsible for education in the regions. They are the spokesmen of the Ministry of National Education in the regions.... At the district levels, there are the Divisional Education Co-ordinators.... At the Ward level there are Ward Education Co-ordinators. After this level, it goes straight to the schools where we find head teachers/headmistresses and teachers....

In nearly all countries there is a minister of education, who may also be responsible for other sectoral activities like culture or sports. This post is nearly always responsible to the national assembly or the council of ministers; in socialist states it is answerable to a party congress and in other cases to a revolutionary council or a supreme body which formulates policy. The government formulates educational policy, at least in its general aspects, and it delegates certain responsibilities to subordinate authorities.

Certain duties and responsibilities generally remain at the highest level but, in accordance with the policy of the country, real delegation of responsibilities to lower authorities always takes place.

The best way of showing how duties are distributed between authorities at the different levels is probably first to show a number of typical examples and then to discuss them. Let us start with a few countries which claim to have decentralized educational administrations.

Sweden has three tiers of educational administration with clear definitions of their responsibilities: the central educational administration, common to the whole country; the regional or county administration; and the local or municipal administration. At the central level there are the Ministry of Education and the National Board of Education with different functions. The former is a political authority which prepares the national education budget, formulates laws, studies educational reforms, makes plans and issues ordinances. The latter is an administrative body under a Director-General, and is probably more important from the point of view of the present book. It interprets educational laws and governmental decisions, and has executive power in this respect. It is responsible for the development of educational content and method, for example, syllabi, curriculum guidelines, regulations about teacher qualifications, educational research and development work, and so on, and it makes sure that resources are used efficiently.

At the regional level there is a county board and inspectors. The board has supervisory and advisory duties; it carries out regional planning and co-ordinates local planning work performed by the municipalities, and it inspects schools. It also organizes some research and development work and, in addition, co-ordinates teacher in-service training and exerts certain control over money paid to the municipalities. At the municipality level there is a local education authority. It is responsible for the use of funds at the local level, for the employment of teachers, and for the building and maintenance of schools. Curricula are established by schools within the framework given by the National Board of Education, and funds are distributed in accordance with agreed regulations. A great deal of authority is vested in local authorities which run the schools. In spite of that, the Swedish primary school system is extremely homogenous. The National Board of Education, working on behalf of the government, issues general guidelines which state the frames of reference within which other authorities may act.

The administrative system in Finland resembles the Swedish one, although it is less decentralized. Decisions on the principles of educational policy are made by Parliament. Educational and cultural affairs are under the Ministry of Education, which is assisted by two advisory bodies for comprehensive and secondary education: the Council for School Affairs, and the Board for Secondary Education Reform. The former body handles all primary school matters. Primary and secondary education are administered by two boards at the national level: the National Board of General Education and the National Board of Vocational Education. The Ministry of Education is responsible for the preparation of legislation and for drafting and implementing political decisions made by the government. The national boards are responsible for the executive administration and the overall planning. Among their other duties are the development of educational curricula.

Under the central government are the provincial governments headed by governors. In each of the eleven provinces there is a school department. The majority of the tasks of these departments concern general and adult education, while vocational education is mainly handled centrally. Local administration is the responsibility of the municipalities, which are self-governing, collect their own taxes and decide certain matters. The decision-making power is with an elected municipal council. In each municipality there is a school council, appointed by the municipal council. Most schools belong to municipalities, but 30 per cent belong to the State, which also meets the major part of the costs for municipal schools.

Colombia has decentralized a number of administrative functions. Public basic education is a national duty. The nation is responsible for costs incurred by the departments, authorities, districts (including the special district of Bogotá) and the whole school system. Teachers are appointed by education secretariats in the districts. Basic education is administered and supervised by these education secretariats, while the Ministry of National Education gives technical assistance to enable them to develop educational policy. Curricula, however, are developed by the ministry and are centralized but flexible. Improvement in and implementation of curricula, the introduction of methodological innovations, the evaluation of the curricula, teacher in-service training, the procurement and distribution of educational materials, the supervision and co-ordination of curricula for non-formal and distance education, and other functions are entrusted to the experimental pilot centres in each district. Co-ordination is the responsibility of a separate division at the Ministry of National Education.

There are many countries which have a centralized educational administration.

Cyprus states that her primary education system is highly centralized. It is headed by a director assisted by two inspectors general and by an inspectorate stationed in four district offices. Appointments, promotions, transfers and disciplinary matters concerning primary education are handled by an independent body appointed by the President of the Republic — the Educational Service Commission. Every school has its own school committee, appointed by the government. Syllabi and timetables are centrally prescribed. Nevertheless, teachers are expected to differentiate the curriculum content according to the special needs of the school.

The Republic of Korea states that there is a *de facto* centralized system. The general objectives and responsibilities of primary education are stated in the Education Law. The administration of primary education is nominally undertaken by the city and provincial boards and offices of education. However, their capacity to manage and administer education is slight, both from a financial point of view and in respect of adjustments to the curriculum. The Ministry of Education is the institution exercising jurisdiction through centralized authority. The local administrative board of education does not have sufficient financial independence to be autonomous. Educational programmes are dependent on funding from the national treasury, and the local boards are formally subsidized by the Ministry of Education, which pays part of salaries and other costs. The ministry develops and directs the administration of the primary education curriculum through the Curriculum Development Centre.

The socialist states in Eastern Europe have a highly centralized educational administration. In the case of the German Democratic Republic, all laws and acts regulating education are passed by the country's supreme body, the People's Chamber. The Council of Ministers is responsible for the centralized planning and management of all institutions of education. It prescribes the content of the work and the structure of the governmental bodies or ministries which are in charge of the planning and running of the individual sectors of education. The so-called secondary schools (ten-year basic schools) are under the Ministry of Education, and, at the regional level, under the competent local education bodies, which have to report to their superior governmental bodies. There is a uniform set of school regulations which governs co-operation between headmasters, teachers, educators, parents, youth organizations, enterprises, trade unions and other groups participating in the education of young people. At the school level, the headmaster of the secondary school co-operates closely with a consultative body — the educational council — and with the parental advisory council. The entire teaching and technical staff is subordinated to the headmaster, who is fully responsible for the running of his school.

The first impression gained from these cases might be that there are great similarities between countries with a very important central administration and a successively less powerful regional and local one. It is hoped, however, that another trend also emerges, namely that in some countries real power is vested in lower-level authorities, power to make important decisions without having to consult higher bodies and without having to obey detailed rules and regulations which turn them into mere executive-level agencies.

The notion of centralization *versus* decentralization should be seen in relative terms: although some systems are more centralized than others, there are always some functions which are under the direct control of a national or — in a few federal nations — a state government. The distribution of power and responsibility takes different forms, however. Generally, overall planning and development of education as well as the drafting of guidelines and the determination of required qualifications for teachers are nearly always handled by the central authority at the national or state level, while the day-to-day administration is a matter for the local administration and the schools. There is a vast area of co-ordination, inspection and interpretation which must be handled by central, intermediate or local authorities. In centralized systems it is handled centrally; in decentralized ones by district or local authorities.

The sharing of authority between the different levels of administration is an important issue and characterizes the education system, for which reason it may be useful to devote some space to this particular point.

In Sweden, as was seen, central authorities issue guidelines and regulations, and handle research and development work of interest to all schools, but important functions are delegated further down the administrative ladder; the county boards thus handle inspection, regional planning and local development projets, while local authorities take care of the curriculum and the employment of teachers. In Australia and the United States, where the systems vary between states, local authorities may even be allowed to determine curricula and do some planning. In Cyprus, on the other hand, nearly all matters are handled by central authorities, for example, appointment, promotion and transfer of teachers. In Uganda, the Ministry of Education provides syllabi and course books, registers teachers and provides education officers, while the development of primary education is the responsibility of committees in each district. Similar systems are reported by China, Ethiopia, France, the United Arab Emirates and other countries.

In Japan, the Ministry of Education, Science and Culture sets forth general guidelines and authorizes textbooks. However, the prefectural governments are responsible for part of the payment of expenditures and the appointment of teachers, while most schools are maintained by local governments. In Malaysia, the Ministry of Education sets minimum standards for schools and provides financial aid, but most schools are run by a headmaster who is responsible to the district education officer. In Egypt, the Ministry of Education draws up plans for education and issues general regulations about curricula, methods of teaching and textbooks, but local governments, in cooperation with district directorates, administer primary schools, establish classes, supervise the application of curricula, fix school calendars, equip schools and so on.

In Poland, the Ministry of Education determines entry and transfer rules for pupils, school location, employment of teachers, examination rules, etc., and maps out policies. The district authorities supervise all schools, while local authorities handle the daily administrative affairs. In Paraguay, the Department of Primary Education in the Ministry of Education co-ordinates the work at the central level: educational policy; administration of human, technical, material and financial resources; operational plans and projects; supervision of educational services; co-ordination of different bodies; and so on. In Cuba, the Ministry of Education carries out and supervises the educational policy and is responsible for: regulations; methodological guidance; standards and procedures; technical advice; training of teachers; research; general and specialized supervision; and so on. The provincial and local offices of education handle the administration of schools and centres and see to it that they function properly. In Iraq, the administration of primary education is the responsibility of the Ministry of Local

Rule (local administration). All funds for this programme are administered at the provincial level by the Director General of Education, who is responsible to the Ministry of Education. All policy decisions, whether short- or long-term ones, are made by the Council of Education, which is chaired by the Minister of Education and includes representatives of all ministries and bodies concerned with the educational process.

Are there any obvious reasons for differences between countries? Although each country presents a separate case and no general explanation of differences can be offered, a few observations can be made on the effects of the general political structure of the nation, on the ways in which funds are collected and distributed, and on historical and geographical factors.

A large number of countries are federations of states, which may enjoy independence in many respects. This means that the central authority is shared in some way between the federal government and the state government. There are several ways of doing this.

Australia is a federal nation. The government of the six states and the Northern Territory have the major responsibility for education, particularly at the compulsory level and including administration and funding of primary education. Each state or territory operates its own system of primary and secondary schools, except the Capital Territory, for which the central Australian Government is directly responsible. There are many non-government schools in the country, usually administered by church authorities, but they must meet government requirements in order to be registered. Generally, schools in Australia enjoy a great deal of autonomy. Most state education departments have established regional administrations, which are responsible for matters like planning, school buildings and deployment of staff, and which are overseen by central authorities. There is often a central curriculum unit which provides general guidelines on course planning, but some systems encourage school-based curriculum development. At the school level, there is considerable community participation in decision making.

In the United States, the responsibility for providing education for children depends on each state, not on the federal government. The operation of the public schools is the responsibility of approximately 15,300 local school districts. The latter have taxing power in most states, lay down basic curricula within state guidelines and determine how resources are expended. Standards for teacher education are generally set by non-governmental accrediting associations. Each state has an educational agency which is responsible for overseeing public and, in some cases, private education. These state agencies may issue teacher certificates and set the minimum curriculum standards. The boards of education that govern the local school districts are usually elected, but in some states they are appointed by a state agency. The boards determine educational policy within state guidelines. They act through a superintendent who serves as chief executive officer and provides leadership to the principals, the teachers and others. The local boards and superintendents must operate within state laws regulating graduation requirements, salaries, health and safety, and so on.

Another federation of fairly independent states is Nigeria where, according to the constitution, primary education is the responsibility of the local government. The report of the Nigerian government to the thirty-ninth session of the International Conference on Education states:

While primary education is financed mainly by the governments of the Federation with the Federal Government contributing the Lion's share, the day-to-day administration is the responsibility of each Local Government with the States Ministries of Education maintaining quality control.

The State Ministries of Education, most of which have special sections dealing with primary education, have the responsibility of evolving appropriate education laws for the State in compliance with the National Education Policy set by the Federal Government. The Ministries are also responsible for inspecting primary schools supplying them with books and stationery; administering them as well as conducting the end of primary education examinations.

In most states, the authority for primary education is decentralized to local education offices in the Local Councils or Local School Boards/Teaching Service Commissions. The Local Governments are responsible for the employment, welfare and discipline of teachers.

Other federal nations organize their primary school systems slightly differently, with the federal governments playing different roles. In Brazil, the federal government looks after education laws, but the organization is otherwise decentralized. In the Federal Republic of Germany, there is a highly decentralized system where the costs of maintaining schools are borne by local authorities while the states *Länder* allocate funds, particularly for buildings and teachers. The teachers are state officials and the schools are state supervised. The curriculum is defined by the ministry of education and cultural affairs in each state, and considerable variations between states are found. There is only one form of primary education, however. In Austria, on the other hand, the basic legislation and the curriculum are under the federal government, while in other respects the primary school system is under the direct competence of the boards of education in the states (*Bundesländer*). Switzerland reports a highly decentralized system with the states (cantons) being responsible for primary education but delegating much responsibility to local authorities. In Mexico too, the system is decentralized. The states are allowed to administer their educational services, but supervision and curricula are said to be reserved for the central government.

A critical point in the decentralization process is often the handling of funds. If the central or district government allocates funds to individual schools, it may exercise a great deal of immediate control, but if the school districts have their own revenues or are given lump sums for distribution, their situation is more flexible.

In most countries, the national government allocates all funds available to public schools. There are other systems, however. In the Federal Republic of Germany, Ireland and Sweden, the state allocates funds for buildings and pays teachers, while in other respects schools are maintained by local authorities from local revenue. In Japan, the national, prefectural and municipal governments share educational expenditures. In Australia, the Federal Republic of Germany, the United States and some other federal nations, federal state and local funds are available, and the states tax their citizens. Many developing countries receive help from international organizations and bilateral aid, and private funds for schools exist in most countries outside Eastern Europe. In this way special needs are met.

The stress given to the different levels of the administrative hierarchy thus varies a great deal. It is likely that part of this variation has its roots in historic development. For example, in the United States, the constitution drawn up two centuries ago made no reference to a federal educational involvement, and therefore the responsibility fell on the shoulders of the states. The same may be the case with other federal nations, although a country like Austria stresses the federal role. It is also likely that small countries, whether measured by population or area, place greater stress on the regional and local educational authorities than large ones, which are more difficult to administer and which are often heterogeneous. This last characteristic may explain why Luxembourg has a highly centralized system where the government determines education in detail with its structure and programmes, trains teachers and so on, while the system in her next-door neighbour, Belgium, is much more decentralized. It may also explain why countries like Israel, Norway and the Netherlands have a weak regional or local organization. On the other hand, a very strong three-tier organization exists in small countries like Sweden and Finland, but weak sub-national authorities in the United Kingdom.

Decentralization of the administrative function is becoming an almost worldwide trend, influenced as much by a call for increased efficiency and adaptation to regional and local conditions as by a will to let new groups of people accept responsibilities for phenomena that concern everybody. Often, however, the distribution of responsibility stops at the district level. Latin America seems to present many examples of this particular trend.

Argentina, which has a rapidly growing population within a vast area, bases her decentralization on a law of 1978 expressed as follows:

The provincial states will have the prime responsibility, within their respective jurisdictions, for the provision, administration and management of the education system at pre-primary, primary and secondary levels.... They will also supervise private education.

In Chile, where the country is divided into three regions, the Minister of Education delegates to regional ministerial secretaries of education functions like the planning, supervision and evaluation of educational activities in accordance with national goals and objectives. There are also municipal education authorities. Countries like Sri Lanka, France, Kuwait and Peru have regionalized a great deal of administration, according to reports.

On the other hand, many countries have gone a step further by strengthening local education authorities. This is the case with Sweden where, as was seen, local authorities even employ teachers, perform local planning and modify curricula. This is also true of the Central African Republic, the Federal Republic of Germany, Guinea, Tunisia, Zimbabwe and other countries, where schools are given greater freedom to decide and act than before. In spite of such changes, education is still a highly centralized system in most countries.

Class organization

Under this heading will be brought together a number of organizational features which may influence the ability of a primary school to provide its pupils with efficient teaching: the grouping of pupils into classes; the length of the school week and year; the school calendar; and the pupil/teacher ratio. These features are determined on the basis of what is considered good for the pupil, and occasionally the teacher, but economic and other frames are also considered. Until recently, many schools in countries which are now highly developed allowed multi-grade classes, that is to say: classes with more than one grade per teacher; part-time education (school only three days per week); a short school year with vacation during harvest time and other 'busy' seasons; and very large classes, sometimes calling for pupil-monitors. These features have now almost entirely disappeared from developed countries as their economic situations have improved and as responsible authorities have become aware of the need for a better class organization. In other parts of the world they still survive.

The effect of class organization is a well-documented educational problem. Studies of the effects of class size and grouping in school are very common [46]. In connection with the Swedish school reform in the 1950s, extensive studies were carried out, for example, by Marklund [44], Svensson [79] and Dahllöf [10]. Many studies are also referred to in the handbooks by Gage [15] and Anderson [2], which have been mentioned before. It can be concluded that the heterogeneity of a class has a negative effect only if it is rather exaggerated, and that factors like the size of the class are important only in rather extreme cases, as when the class exceeds forty pupils.

Although different countries define the concept of primary education in various ways concerning the number of grades and school entrance age, there are great similarities in the way this level of education is structured. All over the world, educators seem to divide primary education into grades which correspond to one year of study. Repetition of grades is often common, but it is always considered an abnormal phenomenon, and non-graded schools are rare.

Another similarity between countries is that the teaching unit considered normal is the single-grade class. This class is also, in principle, age homogeneous, but pupils may have delayed their enrolment in school or may have repeated a grade, which means that age heterogeneity becomes a pronounced phenomenon in many countries. In most countries there are schools which practice so-called multi-grade or split-level class systems, which means that pupils from different grades are taught together in the same room by the same teacher.

Cyprus distinguishes between one-teacher schools, two-teacher schools, three-teacher schools, and so on. In Greece, France and a few other European countries there are many one-teacher schools, and school sizes vary considerably. In Malaysia, there are a few classes with two grades, but in Sri Lanka such combined classes are common in rural areas. In the Republic of Korea, 2 per cent of all classes have more than one grade, usually two, but in India, 35 per cent of all schools have a single teacher and two-thirds of them only two. In Nepal, Thailand, Tonga and Vietnam, multi-grade classes exist. Among the African States, Ethiopia, Madagascar, Rwanda and Zimbabwe report such multi-grade classes, and the Central African Republic mentions that one-teacher schools are common.

All Latin American countries from which information is available report that there are schools with multi-grade classes. Often there is only one teacher per school (*escuelas unitarias*), and this is also the case with the English-speaking countries of Bahamas and Jamaica. Brazil states that multi-grade schools with a single room are very common in rural areas. A common distinction in these countries is made between: complete schools having at least one class in each grade; incomplete schools not providing teaching in all grades; and schools with one (or two) teachers. In the new rural school system in Colombia, age and grade differentiation has been abolished altogether. There are at most three teachers with pupils from all five grades in the same room, working with specially prepared materials. In the Bahamas, there are three types of classes: single-grade classes, combined age-level classes with team teaching, and multi-grade classes. Of the Arab States, Jordan, Morocco and Qatar report two grades in one room if the number of pupils is small.

Countries which exhibit deviations from the system of one grade per class mention that the multi-grade system is prevalent in rural, sparsely populated areas. This is a means for a poor country to extend schooling to remote areas when school buses, boarding schools, distance education or home education have not proved feasible. Multi-grade teaching has been recommended by Unesco missions under such conditions.

Other organizational features which may affect learning in school have also been reported. Australia mentions a number of different types of organization, determined by the individual school. Besides the normal single-grade classes there are several types of multi-grade ones, and in some schools open-plan learning has been adopted with team teaching and several grades working together. Several countries report experiments with team teaching specializing in certain subjects, for example, Peru and Zambia. Open classroom systems are also reported by several countries. The system in Bahamas with combined age-level classes and team teaching in open situations with 50 to 130 pupils was referred to above. The United States reports that in some schools, classrooms may be open so that several instructional groups are in the same room. In

San Marino, the new school system calls for open classrooms and diversification into educational areas within the total space (sports areas, laboratory areas, multi-purpose areas, etc.). This creates a centre for school life where classes and groups (with their teachers) can work together.

China reports three types of establishment: full-time schools; part-time schools which teach only Chinese language, mathematics, general knowledge and moral training; and part-time schools (half-day, every-other-day and mobile schools), which teach only Chinese language and mathematics. Mixed classes with pupils of different grades also exist.

The idea behind the system of one grade/one class is to combine age homogeneity with training level homogeneity. There are, however, variations. Differentiation by ability in sufficiently large schools has been mentioned in a few cases; obviously, this is always the case in special schools and special classes whenever they exist. Ability-homogenous classes as a more general phenomenon are reported by Malta (where the system is said to be practised whenever possible), the Netherlands, Belgium, Jamaica and other countries. Advanced-level classes are found in Czechoslovakia, and classes formed on the basis of academic standard in the Republic of Korea. An age differentiation within the grade is reported for large schools in Brazil. Sweden allows non-graded classes to be organized for grades 1 to 3 on local initiative. The Syrian Arab Republic reports that in areas with few inhabitants a difference of two years from the official age for entry into the system may be permitted. In spite of these exceptions, it must be concluded that the picture emerging from these data is one of relative homogeneity: single-grade classes in densely populated zones and multi-grade classes in areas with scattered populations.

A few of the phenomena which might influence the efficiency of the school systems by determining how much time the pupil spends in school do not, in fact, vary a great deal between countries: the length of the normal school week; the length of the school year; and the school calendar. The normal school week is almost everywhere a little over 20 hours, although it may be divided into periods of different lengths in different countries. Variations between grades may be considerable, however, lower grades having a shorter school week. In Denmark, there is an increase from 18 to 30 periods between grades 1 and 7; in the Federal Republic of Germany and Tunisia from 20 to 27 or 28 between the first and the last grade; and in Sweden from 20 to 34.

The school year seems to vary modestly. Short ones are those in Rwanda (30 weeks) and Turkey (31 to 34 weeks); long ones those in Tonga (42 weeks), San Marino (220 days) and Austria (42 to 43 weeks). Generally speaking, slightly longer school years seem to be found in cool climates than in very hot ones.

The school calendar usually follows a rigid pattern. The school year starts in the early autumn and ends during the early part of the following summer, thus allowing a long summer vacation. Since summer starts at different points of time in different parts of the world, the school may start in February or March in the southern hemisphere, and in August or September in the northern one. Consideration seems rarely to be paid to phenomena other than the season, even though a few countries allow variations. Examples of the latter are Nepal, where the school year usually starts in January but in colder areas in February; and Argentina, where the year stretches between early March and early December, except in cold areas where it is between September and May. In Brazil, Chile, Nigeria and Paraguay, the calendar may be adapted to regional conditions. Colombia has two different calendars. In the northern and central parts of the country, the school year is between February and November; in the southern and western areas it is between September and June. In Turkey, the school year ends in mid-May in rural areas, and June in urban ones. Switzerland has two calendars depending on the canton.

While it is unlikely that these organizational details will have any great effect on outputs from the school system, there are other phenomena which do: the shift school system and the large pupil/teacher ratios found in some countries. Schools with two shifts are reported by many developing countries; multi-shift schools exist in Argentina, Egypt and other countries. Such schools are usually found only in urban areas where lack of buildings and plots to build new schools make it imperative to use existing ones to the utmost. The system has extremely negative effects: the school week becomes shortened, sometimes to only half the normal time, and it becomes impossible to use the schools for extra-curricular activities.

The pupil/teacher ratio varies considerably between countries. Very low ratios, below 20:1, are found mostly in highly developed countries, which are also characterized by low birth rates and aggregated populations, but they are also found in Cuba, where it is said that the ratio will be increased. A few Middle East countries also report fairly low ratios: United Arab Emirates (18:1), Cyprus (21:1), Jordan (22:1), and so on. Very high ratios, over 50:1, are found in most developing countries, for example, Bangladesh, Cameroon, the Central African Republic, Ethiopia, Guinea, Jamaica, Madagascar, Pakistan, the Republic of Korea, Turkey and the United Republic of Tanzania. Very large variations are found within countries, with rural schools usually having smaller classes than urban ones.

Many of the features found in Western schools some time ago can now be found in schools in developing countries, the reason being that the latter countries cannot afford to change them. At the same time, both developed and, to some extent, developing countries try to improve their educational organization by means of innovations like open classrooms, team teaching and so on.

Consequences of the administrative and organizational systems

There are differences between the administrative systems in Member States, but there seems to be a common development in the direction of decentralization. This has consequences for science and technology education. A centralized system means that highly qualified people can be asked to guide the educational development from a central position, that national norms can be developed, that profitable ideas can spread to the whole system, and so on. Decentralization, on the other hand, means (in the ideal case) that regional and local needs and conditions can be taken into account, that people can influence educational development, and that the school can be brought closer to the people. In respect of science and technology, this may mean that courses can be organized taking into consideration local industry, local crafts and traditions. In respect of education in general, it may mean in the long run that people's attitudes towards education change and that schools will start to play an important role in local communities.

The organization of classwork in some countries often shows features which are likely to influence school work negatively. Large schools with one-grade classes — the normal feature in most countries — mean an optimum situation for the teachers, increased resources for acquiring instructional materials, opportunities for teachers to specialize in certain subjects when not all teachers can handle all disciplines, and so on. The situation is less favourable in multi-grade classes, which are usually both badly equipped and badly staffed. Although most poor countries present timetables and calendars very similar to those in richer ones, there are differences, and studies like those by the IEA group [26] indicate that there is a direct relationship between the time devoted to a subject and pupil achievement. The situation is made worse by the fact that many pupils in developing countries are offered inadequate facilities for doing homework and do not receive help from parents; absenteeism is also common. Shift schools are particularly unsatisfactory in this respect. Although a large class might produce results comparable to a small one, increased size makes laboratory work, training in workshops, and other features of modern science and technology teaching less efficient.

EVALUATION OF PUPIL ACHIEVEMENT

Evaluation of pupil achievement in primary school may serve several purposes: (a) to inform teachers and others in the school about the pupil's abilities and performances so as to take the necessary steps for improvement; (b) to inform parents and others outside the school about the pupil's abilities and performance; (c) to inform the pupil so that he/she can improve himself/herself, if necessary; (d) to diagnose the pupil for special placement or treatment; (e) to select the best students for further education or for jobs; (f) to evaluate the curriculum or the teacher in order to make improvements; (g) to study the educational situation, for example, by means of research; and so on. It is rare to find that a clear distinction is made between these aims. In most cases, it seems as if the school systems try to use the same evaluation data for all purposes.

In most Member States for which information about the evaluation procedures is available, the assessment of the pupil is determined by the individual school. There are, however, a number of cases where examinations are organized by a higher body.

In Thailand, the final examination at the end of the primary cycle is set by a district committee, and in Nepal and Nicaragua, it is organized by district offices. In Ethiopia, Mauritius and Zambia, there are national examinations at the end of the cycle, and in Uganda, a primary leaving examination in four subjects in grade 7. Sri Lanka and Bangladesh have competitive examinations at grade 5 to determine those students who will be admitted to 'better' schools and receive scholarships.

Nation-wide examinations at the end of grades 6 and 9 are mentioned by Seychelles. In Malta, an annual examination in grade 3 and higher grades in certain subjects is held on a national level. In the United Republic of Tanzania, there is a primary education leaving examination, which is competitive and prepared by a National Examination Council. In Egypt, the final examination is at the end of grade 6 at the governorate level.

Different techniques of assessing the pupils are reported: observation; questions asked by the teachers during classes; grading of homework; formal written tests and examinations; work samples; and so on. The relative importance of these techniques varies between countries and in some cases between schools. The assessment can have different aims: initial diagnosis; formative evaluation; progress evaluation; and summative evaluation; as well as analyses of the characteristics and qualities of the programmes and the teachers. There is a great variety of measures among different countries: frequent examinations during the year; permanent and continuous evaluation; evaluation at the end of each part of the course (periodic or partial evaluation); mid-term examinations; end-of-term examinations; yearly examinations; final examinations; and so on.

The use of a particular method of evaluating pupils puts a distinctive stamp on the school system. Therefore, a number of examples will be given of ways in which different countries handle the evaluation and examination question. The data are taken from information supplied by Member States to the questionnaire sent out on the occasion of the thirty-ninth session of the International Conference on Education, and the classification of the countries is highly approximate.

Most countries make a great deal of use of examinations. Typically, the examinations are written, even if oral and other types exist, and they are given after each course, at regular intervals during the year, and at the end of the year. There is usually a formalized system of converting examination results into marks.

In Thailand, the different examinations are all handled by the schools. There are pre-tests for initial diagnosis, periodic examinations and final examinations. The latter lead to grade promotion; for this, it is required that the pupil has 80 per cent attendance and achieves objectives for each subject at a certain level. Final examinations in grade 6 are centralized at the district level.

The evaluation of pupils in Nepal is by means of several techniques: observation, homework, unit tests and final examinations. School authorities are free to assign different weights to different evaluation techniques. After completion of primary school, there is a centralized examination.

Jordan also has several types of evaluation: final examinations, mid-term examinations, daily examinations and other activities. They count 50, 25, 15 and 10 per cent, respectively, towards the grade given in each subject.

According to regulations in Nicaragua, all evaluation must be permanent, formative and summative. There are three types of examination: systematic (that is, on-going), partial and term checks. By means of systematic examinations a check is made to see whether course objectives are being met; they can take many forms. Partial examinations, given twice a semester, check whether objectives are met. Term checks measure against semester objectives and are common to a whole district (zone). Partial and term examination scores are added together for the term mark.

In Rwanda, there is continuous evaluation for measuring pupil progress: work during classes, questions, monthly tests and yearly examinations. Promotion depends on the results of these examinations.

In Madagascar, there are periodic and intermediate examinations: monthly, end-of-term and end-of-year tests for promotion from grades 1 to 4. In addition, there are *concours* (competitive examinations) for entry to secondary education from grade 5. The marks are computed by adding examination results together.

In the German Democratic Republic, the progress of the pupil is permanently observed and evaluated by the teacher. Marks are given for oral and written achievement tests, papers prepared at home and classroom work. Marks given during school work are recorded and the pupil is informed. Parents are given reports at regular intervals. Term certificates are based on these marks, as are end-of-school certificates.

The formal and rather rigid nature of the evaluation systems in these countries is obvious, and there are great similarities between the systems. To judge from the country reports, somewhat similar systems are found in Cuba, Egypt, Guinea, Jamaica, Malta, Rwanda, Seychelles, the Syrian Arab Republic, USSR and many other countries. Examination systems of a rather formal nature, but with some deviation from those mentioned above, seem to be found in countries like Argentina, Brazil, the Central African Republic, Colombia, Finland, Mauritius, Mexico, Peru, Poland, Spain and Zambia; in some of these countries the examination burden has been lessened considerably compared to the countries in the previous group.

In Finland, marks are given every term except in grade 1 and in the autumn terms of grades 2 and 3, when they may be replaced by verbal assessments. Marks are based on teacher-constructed and administered tests; certain marks are required for transfer to a higher grade, and poor grades may call for a new examination.

Mexico conducts continuous evaluation by means of systematic observations and occasional tests and exercises. Periodic evaluations use data collected during the continuous evaluation and by means of teacher ratings at the end of each course unit. Final evaluations are made on the basis of the partial ones and the teacher's judgement of pupil development in different curriculum areas.

The evaluation in Peru should be flexible, integrated and on-going. It aims to measure pupil achievement, stimulate the pupil and provide information. It uses diagnostic, initial, periodic and final examinations, which form the basis for grade-level certificates.

Zambia makes continuous evaluation through observations and periodic tests given by the teachers. At the end of primary school, there is an examination by the Ministry of Education and Culture.

According to the data made available by Member States, examination systems in some countries could be characterized as more-or-less independent of formal tests and examinations, even if teacher observations and occasional informal examinations play a large part; this is true of Austria, Belgium, Bulgaria, Cyprus, Denmark, Federal Republic of Germany, Ireland, Japan, San Marino, Sri Lanka, Sweden, Tonga and others.

In Bahamas, teachers keep continuous records for each term. The evaluation is school based, but plans exist for national instruments for grades 3 and 6; at present, there is a placement test in grade 6.

Sweden has continuous evaluation during primary school, usually without standardized tests. The teachers present their evaluations of the pupils' achievements during individual discussions with parents. There are no marks in grades 1 to 6.

In Bulgaria, teachers must evaluate the pupil's knowledge, aptitudes and habits but do not do this in numerical terms, and the evaluation must have an educational effect. Parents are periodically informed about their children's performance. No report cards are presented except for the end-of-year certificate.

The standard evaluation method in Tonga is a final examination given at the end of primary education for entry to secondary school. No certificate is given but a progress record of each pupil is kept in the school files, and each teacher keeps records for assessment purposes. Periodic evaluation is made throughout the school year, but the method varies between schools.

There are thus great variations between those countries like Thailand and Nepal, which use many kinds of formal instruments repeatedly during the year, and those like Bulgaria and Sweden where formal grading plays a small part and reports in forms other than marks dominate. The use of formal examination instruments is defended by many educators on the grounds that: they provide a means for safeguarding the quality of the system; they enable placement of pupils; they form a suitable way for reporting pupil progress to parents, school administrators and others; and they encourage pupils to work. Critics of examinations point to weaknesses: they place the emphasis on memorization and knowledge while modern schools stress understanding, application in new areas and creativity; they direct the teaching towards examination criteria rather than meeting educational objectives; they disrupt the work in school; and they take up a lot of time. Furthermore, educational measurements like examinations and marks suffer from low reliability and often from low validity [80, 89]. It seems obvious that both types of arguments are, in principle, based on facts; the use of formal assessment systems with oral and written examinations leading to marking has advantages as well as disadvantages. The important question is whether the former are more important than the latter. Many countries, particularly so-called Western ones, are using examinations and marks less and less.

Most countries report that certificates are awarded at the end of primary school. This usually also means that the pupil is accepted for further education. In a number of countries there is no certificate at all, for example, in Australia, Austria, Cyprus, Czechoslovakia, Denmark, Finland, the Federal Republic of Germany, Malta, Pakistan, the Netherlands, Sweden, Tonga and the United Kingdom. In Colombia, there is no certificate, but a letter of assessment is issued. Sri Lanka has no certificate but holds competitive examinations for entry to the next higher level of education. In Egypt, those who succeed in basic education are given the Basic Education Certificate; those who fail receive an attendance certificate. A certificate that does not give access to further education is reported by the Central African Republic and the United Republic of Tanzania; in both cases there is a further selection of applicants. In Algeria a certificate is given at the end of compulsory education.

TEACHER TRAINING

Teaching is a long-standing profession, but systematic teacher training is a comparatively recent phenomenon and has hardly assumed its final form. There are still countries where future teachers are given very little training. With the development of education as a science and with the general realization that teaching skills can be improved thanks to training and that knowledge of child psychology and other basic knowledge is important to teaching, the emphasis on prolonged teacher training has become accentuated. Nowadays, the trend is definitely to require a specialized university degree, even for primary teaching. Another trend worth observing is that training for different kinds of teacher has become more and more uniform. While teachers were previously prepared as specialists in their disciplines or for teaching at different levels, all teachers now tend to have a broad base of similar training in 'education'.

The effect of teacher qualifications is a common area for research. Studies by, for example, Chaïbderraine [4] and the World Bank [11], as well as summaries presented by Guthrie [18], Rosenshine [68] and Simmons [74] indicate a positive relationship between teacher variables and pupil achievement. In Chaïbderraine's case, it was shown that completely unqualified teachers do not perform as well as others; in other studies it was shown that teacher behaviour, influenced by teacher training or teacher experience, has a positive effect on pupil performance in primary school.

The following discussion is essentially based on responses to the International Bureau of Education's pre-conference questionnaire about the introduction to science and technology in primary school. It is evident that the training of primary school-teachers varies a great deal between countries, and in some cases there seems to be limited pedagogical training. Space does not permit a complete discussion of the data given; only a few general trends will be illustrated.

The backgrounds of primary school-teachers

Many countries require completed secondary school and long subsequent training at a university, a teacher training centre or a 'normal' school before a teacher is considered qualified. In Finland training lasts for four to five years; in Denmark, Greece, Mexico and Zimbabwe four; in Chile, Czechoslovakia and the United Kingdom three to four; and in the Federal Republic of Germany three years plus a one-year period of teaching practice after secondary school graduation. In quite a number of countries this post-secondary training takes about three years, for example, in Australia, Cyprus, Guinea, Hungary, Ireland, Jamaica, Luxembourg, Malaysia and Sweden. Two to four years is reported by Argentina, the Republic of Korea and Saudi Arabia; two years plus sixteen weeks of practice by Belgium. There are a few countries which require only short pre-service training at university-level institutions: two years by France, Turkey, Uganda and the United Arab Emirates; one to two years by Algeria and Sri Lanka; one year by Ethiopia; and six months by Nepal. In a number of countries, the teacher training leads to a first-level academic degree.

The delivery system

In many countries, the entry requirement for teacher training is graduation from intermediate (lower secondary) school, essentially nine or ten years of schooling. This is the case with, for example, Egypt (five years of teacher training), Nicaragua (three years of teacher training), Bangladesh (one or two years of teacher training), Madagascar (where basic secondary school training and elementary pedagogy are given at teacher training institutes), Bahamas, Guinea and Seychelles (where the teacher training is for three years), the United Republic of Tanzania (two years of teacher training), Jamaica (where the 'old' system means two years of teacher training plus one year of teaching practice), Pakistan (one year of training), and the Netherlands. In Brazil, teachers for grades 1 to 4 have this background. Even lower background requirements can be found in some cases.

Parallel systems exist in many countries. Cameroon reports not less than three levels of teacher training, that is, streams with different entry requirements. In Switzerland, the entry requirement may be nine years of schooling or secondary school graduation. In Uganda there are several levels. China accepts secondary school, senior middle school and junior middle school graduates. In a number of cases, teacher trainees with different entry backgrounds receive different qualifications.

The employment of many unqualified teachers is reported by several countries. In a few cases, the countries try to remedy this situation by means of systematic in-service training. On the whole, the question of teacher qualifications and teacher training is very haphazard. Different kinds of teachers often work side by side, as a result of parallel systems or because the teacher training system has changed. The staff available in the teaching field is composed of teachers who have been trained under various systems over a period of at least forty years. While the requirements for entry into teaching careers have become higher and higher, many long-serving teachers may have received training that is inadequate or non-existent.

Unesco made a study of the qualifications of primary-school teachers in a number of countries about ten years ago [86]. While such training is almost the rule at present, teachers with third-level training were then rare in many countries. Teacher training at the upper-secondary level was still common. While few countries nowadays provide teacher training at the lower-secondary level, it was quite usual until recently and there could still be many such teachers in certain countries, which suggests a considerable weakness in the system.

While secondary school-teachers are always or almost always specialists, those in primary school are more often generalists, able to teach all or almost all subjects. There are, however, a few subjects which are usually taught by specialists since the generalist teacher cannot be expected to master them. This is reported by, for example, Japan, where some practical subjects are often taught by specialists; Nepal, where a few semi-specialists exist; Tunisia, where there are specialists in manual work; Mauritius, which has specialists in oriental languages, and so on. Other particular examples are religious instruction, art, physical education, music, home economics, sewing and modern languages, all reported to need specialist teachers by several countries.

Teachers specializing in some general subjects are reported by a few countries. Cuba reports several broad areas of specialization: mathematics, Spanish and history, biology and geography, work and artistic education, and pre-school education. In Chile, there are specialized subject teachers in the upper grades. In Spain, there are semi-specialists in broad areas: social sciences, mother tongue, foreign languages and pre-school education. In Ethiopia, subject teachers take over from grade 4. In the German Democratic Republic, teachers at the lower level of primary school are qualified to teach German, mathematics and all optional subjects, while those at the higher level are subject specialists. In Bulgaria, classwork in primary school is reported to be divided between teachers according to their qualifications.

As already mentioned, in some countries there are one-teacher schools with teachers teaching not only all subjects but all grades as well.

The training of science and technology teachers

Pre-service training of teachers of science and technology presents a rather homogeneous picture. Most countries report that general teacher training is considered an adequate preparation. A typical example is given by Australia, which states that pre-service education takes into account the latest trends in primary education and that there is professional liaison between teacher education and educational authorities in respect of curriculum development. Uganda describes a situation where teacher training is planned in relation to the curriculum, and student teachers are trained in the use of the new curriculum material.

In a number of cases it is reported that some specialization is provided during training. Algeria mentions that the introduction of the polytechnical school is accompanied by a reform in the profile and content of teacher training, and new streams for science have been created. In some areas of the United States, there are teachers specializing in science and mathematics, but this is still rare at the primary level. Quite a few countries report that, while teachers in the lower grades have a common background of teacher training, those in the upper grades have received special training in some subjects. In Chile, for example, teachers in grades 7 and 8 receive specialized training in subjects like mathematics; in Seychelles the same is true of teachers in grades 7 to 9; and in the German Democratic Republic with those in grades 5 to 10. In the new eight-year school in Turkey, science teachers need university training in their subjects. Tunisia reports that certain technology teachers receive supplementary technical training to be able to teach in grades 5 to 7, while Turkey prepares technology teachers in faculties for professional and technical training. In many cases, technicians and craftsmen are used as teachers or instructors of vocational subjects in primary school.

In-service teacher training is reported to be the main means for promoting science and technology. The vast majority of teachers in many countries did not receive the type of pre-service training that would enable them to teach new or restructured subjects of this kind, and it takes a long time before the influx from pre-service training can change the situation. In-service training as a pre-requisite for science and technology education is therefore mentioned by a few countries.

In Nigeria, these subjects are taught by the same teachers as other subjects, but the teachers receive in-service training in workshops. The United Kingdom states that science and technology teachers have been given initial training, but the vast majority of class teachers will further develop their skills through in-service and in-school training. In Sri Lanka, in-service training is the main mode for the preparation of teachers for the new curriculum. Of those already serving, a few follow courses to become teacher-educators who can conduct in-service training in schools. In Jordan, there are two forms of teacher education in science and technology: attendance at community colleges combined with in-service training; and in-service training alone.

Not much information is provided about the nature of the in-service training given in different countries, but most countries report simply that such courses exist. Paraguay, however, mentions that primary school-teachers participate during the school year in courses, study days and pedagogical workshops. The Republic of Korea organizes annual courses using special material. Guyana uses an emergency science programme to update the skills of practising teachers. In Uganda, in-service courses are organized and existing teachers' centres are being revived. In Algeria, the new polytechnical school is introduced by means of in-service training, seminars and study days with demonstration of methods, and workshops. Vacation courses are mentioned by, for example, Brazil and Peru; short courses by several countries. Study circles in schools, workshops, study days, etc., are seemingly common phenomena. In Viet Nam, schools should be linked to social life, and most of them are supported by local scientific and technical institutions. This means that rural teachers are given further training in agricultural technology by local organizations, while urban teachers are provided training in industrial fields. Many different organizers of in-service training are reported: the ministry in question or one of its departments; special centres, universities, colleges and teacher-training schools; teachers' associations; supervisors; various governmental and private organizations; and the schools themselves.

In-service training is a haphazard affair in most countries; for daily help and advice to teachers, other means must be available. Many countries mention publications like handbooks, textbooks, guides, periodicals, technical circulars and documents about content and methodology. Science kits are mentioned by the Bahamas and Nigeria; satellite instructional television by India. Detailed printed information about the system is given in Hungary, where relevant research information has also been distributed. Ireland, Jamaica and the Netherlands describe school libraries equipped as teachers' resource centres, and Switzerland refers to centres for education and documentation, which are open to teachers. Cameroon mentions co-operation between the schools and local artisans. Other fruitful ideas include societies for nature study (China), orientation courses on new curricula (Uganda), encouragement to teachers to develop teaching/learning materials themselves (Japan), and attendance at classes run by well-trained teachers (Cyprus).

CONSEQUENCES FOR SCIENCE AND TECHNOLOGY EDUCATION

The preceding discussion has dealt with important characteristics of the delivery systems in different countries. It has not always been possible to distinguish between what is relevant to science and technology and what applies to other educational areas, although an attempt has been made and the stress is placed on the former subjects. It is felt that the delivery system is often the same for the whole curriculum content and that general weaknesses and strong points would usually affect all areas. Science and technology education presents specific problems, however, which will be pointed out below.

Four areas of the delivery system have been found particularly important and therefore discussed in some depth: teaching methods; administration and class organization; evaluation of pupil performance; and teacher training.

All over the world it seems generally accepted that the teaching of science and technology shall aim to create an understanding of relationships and an ability to solve problems. Various skills are also stressed, while memorization and the learning of isolated facts are considered less important. This has led to classroom procedures like group and individualized work, use of laboratory methods and discovery, and a scientific approach in general. The teacher's role has been changed from one of delivery of material to be learned and explanations to be absorbed to one of organization, co-ordination and participation.

Such a change in the teaching situation places specific demands on the school system. Memorization is becoming less important, but as a consequence, more stress must be placed on the availability of source materials, like books and other written materials, pictures and samples from the field studied. Instructional materials and physical facilities must be available for the purpose, which means that increased demands for well-equipped schools appear. The schools must be administered and organized for the purpose, which means that it should be possible to adapt the curriculum to local conditions, that sufficient time should be devoted to each subject, that teachers should have sufficiently small classes, and so on. The examination system must be changed to fit the new ideas. The teacher must be adequately trained for his new tasks. There is grave doubt whether these prerequisites are always met.

In poor countries, very little money is usually available for the procurement of books and equipment. There are countries where the dearth of reading materials causes grave concern, as when books are available only in a few subjects or to a few pupils. Even when well-trained teachers can create their own teaching materials and other materials may be available locally, efficient instructional resources like television programmes, science kits and libraries cost a great deal, and the classes are furthermore often cut off from modern technological developments involving, for example, computers. Poor countries also lack science and technology stimulation in the environment and the home.

Earlier, a review was made of whether the educational organization in different countries can be classified as centralized or decentralized. Most countries seem to agree that decentralization is essential, one reason being that it becomes easier to adapt curricula to local conditions. There is a long-term trend towards decentralization, but most systems are still highly centralized. Where decentralization has taken place, it is often only to the regional level, while local school authorities are rarely given any real power. Exceptions are mostly found in the Western world.

Among features on class organization, the chapter raised the problem of single-grade and multi-grade classes, the school week and the school year, and the pupil/teacher ratio. Multi-grade classes may provide an excellent educational climate for children if the schools are well equipped and the teachers well trained. Unfortunately, such classes are mostly found in rural regions in poor countries, often in areas where other factors like geography, parental attitudes towards education, the need for the children to work, etc., have a negative influence on education. Teachers in multi-grade classes are rarely well trained for this situation, and in many cases they have elected to go to such a school because they could not find work elsewhere.

The length of the school week and the school year do not vary very much, but they are slightly shorter in most developing countries compared to the majority of developed ones. Features like shift schools and prolonged absenteeism for other than health reasons are practically unknown in developed countries. Although small pupil/teacher ratios can be found in a few developing countries, it is much more common to find that they are extremely large. This means a heavy stress on the teacher, who would be inclined to rely on traditional teacher-directed modes of teaching, which are also often encouraged by the social environment since they are perceived as the 'right' ones.

Examinations are not necessarily a bad phenomenon, but traditional written ones have definite disadvantages from the point of view of modern science and technology teaching: they take up a lot of time, and they divert the attention from comprehension in favour of memorization and knowledge of isolated facts. Many Member States, mainly developed ones, have sought other ways of assessing and regulating pupils, but in most developing countries heavy stress is still placed on formal examinations.

The trend in teacher education is towards university-level training in a subject, as well as pedagogy, psychology and teaching methodology. However, many teachers, particularly in Third World countries, still have an inadequate background with short theoretical schooling and even less pedagogical training. Some teachers have little more than primary education themselves, and many are unqualified. In many developing countries and in a few developed ones as well, the science and technology background of those who enter a teaching career is very meagre, and the scarcity of written materials and science kits in the schools makes it difficult for them to improve themselves. Their environment is usually as poor in science and technology as that of their pupils. With such a background, in-service training is faced with an even greater challenge, so that teachers will find it hard to give the young generation a scientific and technical education for an adult world that is becoming more and more complicated.

What has been said does not necessarily mean that education in developing countries always presents a less promising picture than in developed ones. In the first place, the preceding discussion concerns primarily science and technology education, where, quite naturally, developed countries hold the upper hand. Secondly, the situation varies between groups of countries and particularly within countries. The present authors have seen developed countries with a number of very bad schools, as well as developing ones with excellent science and technology education. The school systems in poor countries would yet need more support than those in rich ones to develop their science and technology education.

In the introduction to the present chapter, reference was made to the 'frames' which surround an educational activity. In the subsequent discussion it has been found that certain frames influence the ability of a school system to provide good science and technology education, and the following ones have been stressed: availability of money, equipment, books and other reading materials, etc.; the skills and backgrounds of the teachers; the administrative and organizational structures; attitudes in different social groups; and so on. To improve the situation must mean to widen the frames: to provide more resources, to train more qualified teachers, and so forth. This will be discussed in the next chapter.

CHAPTER VI
Renewal of science and technology education

There is a proverb in many languages implying that to stand still means to take a step backwards. The surrounding world is in constant development and a person, or a nation, that does not participate in this development will soon be left behind. This is certainly true of the field of science and technology, where discoveries, inventions and developments not only seem to appear at an accelerating rate but also play a larger and larger role. Let us recall statements which have been made before: to improve economically and in other ways a country must participate in the scientific and technological development process and, to be able to do so, its education system must be alert and must renew itself in accordance with current needs.

There are also other reasons for educational renewal. In many countries it is found that primary education does not function efficiently, with well-known effects such as wastage, inequities and high costs. It is felt that this situation must be changed. In other cases the system does not make use of educational and psychological advances in teaching techniques, and this should be remedied. In still other cases, useful philosophies or ideologies have appeared which should influence thinking within the educational sphere. Whatever the reason, most countries experience a need for educational renewal.

Educational authorities, politicians and educators in all countries are certainly aware of such needs, and educational innovations are constantly being introduced. This chapter will bring up some aspects of the renewal of primary school. Emphasis will be placed on changes affecting the teaching of science and technology.

The preceding chapters have dealt with different aspects of the educational process in school: educational goals and objectives, the structure and content of the curriculum, and the delivery system. The renewal process may touch upon any or all of these aspects. The following discussion will describe examples of what is going on. However, what is a new and exotic dish in one household may be looked upon as warmed-up left-overs in another; countries are at different levels of development, and much of the renewal taking place is simply well-established ideas being applied in new circumstances or in new countries.

Renewal processes do not simply happen; they must be prepared for in many ways. The introduction of an innovation causes a series of changes of a theoretical and practical nature; they mean a step taken into an often badly known area. Administrators, teachers and others may not know how to handle the new situation. Therefore, this calls for preparation by means of educational research and teacher training. It is likely that the renewal has a number of effects, even apart from the expected one of improving education in the area directly concerned, and such effects — positive or negative — should be considered. Like other processes in an education system, a renewal is influenced by the 'frames' which limit it or sometimes facilitate it. Frames were touched upon in the previous chapter; they will be more thoroughly discussed below.

Innovations and renewal efforts have been mentioned before in this book. In Chapter II there was a discussion of steps taken to bring children to school and to reduce wastage in education. In the following three chapters, reference was made to changes that have been made in some countries to express more relevant goals and objectives, to modernize curriculum content and structure, or to improve the educational delivery system. Below, however, there will be a discussion of the current and planned renewal of primary education, based on the reports sent in by Member States. Although efforts have been made to avoid duplication, it is inevitable that overlapping will occur and that reference will be made to renewal programmes that have already been described.

THE GOALS AND OBJECTIVES FOR RENEWAL

Educational aims or, better labelled, goals and objectives can be of many kinds. In Chapter III, they were divided into: unspecified goals; ideological, cultural and political goals and objectives; educational goals and objectives; goals and objectives of person-centred change; and goals and objectives of societal change. The aims of renewal can also be classified into these categories, and a comparison with those of the existing education system may prove of interest.

Goals and objectives for the renewal of primary education

Not all renewal processes are of equal importance. In some cases the education systems are said to function well, while in others they may be in need of a complete revision. A few countries which answered the International Bureau of Education's questionnaire about introducing science and technology stated that their systems are undergoing consolidation rather than a renewal process. This is the case with, for example, Austria, Brazil, Denmark, the Federal Republic of Germany, Sweden and the United Kingdom. Austria continues a renewal process which started in 1962. Denmark is said to have passed through a long process of reform involving all aspects of the primary education system, and 'nothing spectacular' takes place. Brazil reports that great progress has been made as a result of renewal efforts. The Federal Republic of Germany has gone through a long reform process, and in Sweden most people seem to want a period of calm after reforms which have modified the whole education system in fundamental ways. In the United Kingdom, the history of renewal is reported to be one of a well-established partnership between teachers, local and central organizations, and professional associations.

In some countries educational renewal cannot be perceived as a separate process, but the system does change continuously.

Guinea reports such a process which has been going on since 1959. Poland describes a primary education system undergoing continuous change. In China, curricula and materials have been adjusted at irregular intervals to improve their relevance. The United States reports that no national renewal takes place, but in every state there are task forces concerned with educational reform.

Quite a few countries strive for a renewal with *ideological, cultural or political* aims.

Nicaragua, having undergone radical political changes, plans a complete transformation of her whole education system. This was initiated with a national campaign against illiteracy. The main objective is to give students an education which makes them see themselves as agents for self-education and enables them to reach high educational levels. A determination of goals and objectives for what is called 'new education' is going on. This education shall educate fully and in an integrated way the 'new man', able to contribute to building the 'new society'. In Rwanda, the renewal effort aims for democratization with compulsory education and for socio-economic and cultural development. Madagascar states that her education shall be in line with her political development. The Republic of Korea mentions a re-evaluation of her educational policy, and Egypt stresses the goal of preparing good, capable citizens. Australia and the Netherlands find it necessary to recognize that their societies are multi-national and multi-cultural.

A small number of countries mention the need to adapt to new political ideologies which value all kinds of work.

In Congo the 'school for the people' has been given completely new curricula and a new organization in order to bring the school culture into close contact with workers and production. India and Cameroon, like European socialist countries, stress the political goal of uniting intellectual and work training. The German Democratic Republic wants education to reflect the spirit of peace and disarmament, international understanding, solidarity and the implementation of human rights.

The ideological, cultural and political aims of education are obviously important, and most countries stress them in their statements of goals, as reported in Chapter III. That few countries bring up such aims in respect of educational renewal may seem surprising, but this may mean that such goals and objectives are usually well established and that little need is therefore felt for giving them further emphasis.

A great many educational goals and objectives for renewal are expressed by Member States.

The need for an adaptation of primary education to national, regional or local conditions is expressed by many countries. Belgium, Bulgaria, the Central African Republic and Jordan aim to adapt their education to national realities and integrate their schools into the environment. One aim of the Republic of Korea is for a curriculum related to radically changing conditions in society. Seychelles expects that learning shall be related to life and work and be consistent with national developmental goals. Senegal wants adaptation to national reality through the introduction of national languages, the progressive opening of environmental schools and the introduction of productive work. Colombia takes into consideration the special characteristics of regions in the country, and Finland prepares curricula reflecting local characteristics.

Bringing the education system into line with changes that have occurred in society and in science and technology is emphasized by Australia, the Central African Republic, Cyprus, Ethiopia, the German Democratic Republic, Mexico, Nigeria, Poland, Spain, Sri Lanka, Turkey and other countries. Jordan and Sweden point to the need for adapting education to developments and changes in society and working life; France to the need for adjusting to changes in the cultural environment, such as the spread of new media.

That the quality of the education given is felt to be inadequate is of major concern in a series of Member States.

The Republic of Korea questions the validity of curriculum content in the primary school. Mexico and Paraguay state that their education does not always meet individual and societal needs. Paraguay, Tunisia and Zambia point to wastage and other aspects of an inadequate system, and Tunisia also states that too many pupils fail in primary school or leave without proper preparation for work, which leads to demands for better quality. Pakistan wants to improve the quality of learning and cut costs, and the Byelorussian SSR and Ethiopia feel a need for improved quality in all areas: content, materials and methods. The Bahamas stress reinforcement of the quality of primary education since it is felt that many graduates do not master basic subjects. On the other hand, there are several countries which intend to expand education quantitatively, for example, Rwanda, which wants education to reach the broad masses. The Republic of Korea aims for a balance between quantitative and qualitative development.

For obvious reasons, much of the renewal concerns the content of the curriculum, which shall be brought into line with developments.

Several Member States specify particular areas in which there should be changes in content. Aesthetic and practical subjects and physical activities are mentioned by France and Finland, practical work by Rwanda, basic skills by Australia, creative activities by Sri Lanka, 'new maths' and other subjects by Tonga, and so on. Cameroon wants to unite intellectual and work training, as well as integrating the former French and British education systems in the country. Uganda aims to broaden primary education to include cultural and academic components and marketable skills. Switzerland foresees a profound revision of the teaching of certain subjects, new curriculum areas and the teaching of a second language. Qatar aims for a science-centred curriculum. India's renewal effort includes an innovative curriculum with instructional materials relevant to different groups of children. A special case is presented by the United Arab Emirates, which had previously adopted the curriculum of Kuwait but will now build up a national one including vocational education.

Changed teacher training to meet the needs of the new curriculum is mentioned by several countries, including Pakistan, the Republic of Korea, Qatar and the United Republic of Tanzania. Decentralization with curricula prepared by local and regional authorities is planned by Argentina, Finland and others; less detailed curricula by Australia; and an adaptation to local and regional contexts by Finland. Other potential organizational changes include: a new curriculum examination board (Ireland); improving the situation for intellectually or otherwise handicapped children (Malta); and new approaches for dealing with handicapped children (the Netherlands). Open classrooms and open-space buildings are mentioned by several States. Iraq plans coordination and educational guidance for head-teachers and school libraries. Pakistan enumerates new facilities, materials, equipment and supervision amongst other changes. The Republic of Korea stresses the importance of better administrative support, and Thailand mentions administrative changes.

A number of countries aim for a renewal which will bring their primary education into line with modern pedagogical thought and with the outcomes of educational research. This is expressed in these terms by Cyprus, while San Marino mentions moving away from teaching aimed at purely didactic goals towards an education which takes wider ideas into consideration. The German Democratic Republic states that all innovations shall be closely linked with the results of scientific research and practical improvements made by teachers. Malaysia found that the old curriculum content was overloaded and therefore a more basic one has been introduced; and Japan tries to remedy the learning overburden so as to contribute to the development of children who are intellectually and physically balanced. Rwanda states that the primary education reform has changed the quality of the education system, which was previously elitist; and Czechoslovakia wants to enable children to progress through the system, and therefore concentrates on essentials.

The other major group of goals behind educational renewal in Member States are those of *person-centred change*. In most cases these goals imply that primary education shall be adapted to the child's characteristics and situation.

This is stated differently by different countries. The Republic of Korea wants education to reflect the abilities and aptitudes of pupils by individualizing learning. Uganda tries to create an education relevant to the learner, to relate the pupil to his environment, and to make the course content relevant to his/her experience. India adjusts curricula to the children's life styles and to socio-economic opportunities. Mention is made of flexible, problem-centred, work-based and decentralized curricula relevant to the objectives of national development and to the needs and life situations of different groups of children. The Netherlands states that school shall be geared towards enabling children to undergo an uninterrupted development — emotionally, cognitively and creatively — and to learn social, cultural and physical skills. The Syrian Arab Republic stresses the balance between physical, intellectual, social and psychological development. Seychelles wants the school to develop self-reliant people; Jordan to help pupils develop their attitudes and abilities; and the German Democratic Republic to exploit all reserves with regard to personality development in the lower grades. Algeria, the Federal Republic of Germany and Tunisia state the aim of improving the ability of the school to do justice to the aptitudes and interests of the individual child. Chile states the goal of an education which considers the characteristics and needs of the learner, his family, his community and available resources. Peru wants to consider the dignity of Man and the importance of the individual.

The fourth quadrant of the model (see Figure 3, Chapter III) covers the goals and objectives of *societal change*.

Such goals behind the renewal have been mentioned by a few countries. Improved quality of educational opportunities is mentioned by the Federal Republic of Germany, Paraguay, Peru, Poland, the Republic of Korea, Seychelles and others. Expansion of primary education and the re-introduction of compulsory education is a renewal goal in Jamaica. Malta plans special classes for low achievers and handicapped children, and the Netherlands mentions new approaches for dealing with handicapped pupils. Pakistan wants to expand enrolment, primarily of girls, while many countries feel the need for a prolongation of basic education since the existing compulsory system has proved inadequate. Egypt wants a system which enables children to retain literacy. The Syrian Arab Republic states that primary education shall answer to the aspirations of educational policy and be able to match the progress of teaching and socio-economic progress in the world. It shall educate an Arab citizen who can influence social and economic conditions in order to bring about improvements in society.

Reading the statements made by Member States about the goals and objectives of their educational renewal gives a clear impression that they aim primarily to change the curricula in accordance with developments already taking place within or outside the country, and to adapt them to the child's nature and situation. These are person-centred goals, as discussed in Chapter III. Goals of cultural, ideological and political developments, and goals of societal change are mentioned to a much smaller extent.

Purposes for the renewal of science and technology education

The data on goals and objectives shown in the above section concern primary education as a whole. Science and technology education obviously share these goals. There might, however, be cases when the specific character and possibilities of these subjects cause particular aims to be stressed, and in other cases certain goals might only be met by means of these disciplines. For this reason, it is necessary to study the goals of science and technology education in isolation.

The questionnaire sent to Member States in 1983 contained a section about criteria behind the renewal of science and technology education. Most of the countries interpreted it as dealing with goals and objective. Although care must obviously be taken when discussing the data, particularly when making comparison with other information about educational purposes, the answers to this question provide important information about the ways in which countries want their science and technology education to develop.

Only a few countries have made reference to the fact that the renewal of science and technology education also has *ideological, cultural or political goals*.

Hungary lists a series of general criteria behind science and technology development, among which are the present and long-term requirements of society, social culture and a scientific world view. The Netherlands state the aim of giving children an adequate introduction to general and specific aspects of culture, and further refer to the goal of giving young people opportunities to find meaningful employment in society. It is pointed out that scientific development and the use of technology form a considerable part of our culture and it is necessary to familiarize children with them from an early age. Chile also mentions science and technology in connection with cultural goals. Viet Nam emphasizes the close connection between the teaching of science and technology and the education of 'socialist man'. Kuwait discusses adaptation to modern points of view, and the United Arab Emirates the connection between educational policy and societal needs.

Most countries declare that the renewal of science and technology education is guided by *educational* goals and objectives; in other words, that the development of science and technology shall be in accordance with the stated educational purposes and policies of the country. Chile, Jordan and Nigeria modify the content and structure of the curriculum in relation to the development of their societies and the requirements of social change. The programmes for science and technology education in Turkey take into consideration changes in science, technology and economy. The German Democratic Republic and Cuba stress the polytechnical nature of socialist education. Zimbabwe finds that curricula must be developed to fit the general economic and environmental situation of the country, and Paraguay intends to place the focus on the rural environment, on problems actually found, on relevant solutions and on regional necessities. Cuba intends to integrate science and technology curricula with those of other subjects.

That primary education shall be relevant to science and technology development is emphasized by Senegal and the United Republic of Tanzania; Hungary mentions scientific thinking; and Australia states that the main criteria behind the development of science and technology education are the development of skills leading to an understanding of the nature of scientific inquiry, and the equipping of the pupils for a responsible and appreciative interaction with their environment. The Republic of Korea lists the goals of acquainting pupils with nature and fostering positive attitudes towards life, and India mentions the goal of changing curricula and the delivery system in accordance with the rapid rate of technological development. In Czechoslovakia, a criterion in the plan for science and technology teaching is to concentrate on the importance of the natural sciences. The relationship between science and technology curricula at different levels of education is found important by Algeria, Pakistan and Switzerland.

A curriculum developed in accordance with the results of scientific research is a goal in countries like France, Hungary, Japan, Jordan, Mauritius and Nigeria. Several ways of doing this are suggested. Nigeria, for example, develops curricula as a result of the deliberations of a panel of specialists, while in France the Institut national de recherche pédagogique assembles teacher trainers, school inspectors and educational advisers to study development. Jordan applies the results of educational research on science and technology teaching and those of psychological research on the teaching/learning process. Mauritius wants tools of inquiry to be taught in the classroom rather than simply inculcating knowledge. Hungary stresses the logic of educational processes and considers the system of science, its present level of development and future trends.

A basic principle in the renewal of science and technology education in many countries is that regional and local conditions shall be taken into consideration.

Turkey and Japan develop curricula which vary according to regional and local needs, but within a nationally determined framework. In Madagascar, curriculum projects are developed at the local level by teachers and administrators; in Finland and Sweden, local authorities and schools are responsible for this; and in Zimbabwe, there is some room for teachers to modify methods and content according to set objectives. Argentina reports that curriculum preparation is decentralized with the participation of teams of teachers and technicians who take into consideration educational philosophy, the features of the region and the community, and the characteristics of the children. In China, the curriculum outline for science and technology education is drawn up by local bodies; and in Colombia, initial suggestions at the central level are further developed by teachers, pupils, associations and the community in general.

Nearly all countries which express criteria for the development of science and technology education mention the *nature of the child and the environment*.

The vast majority of countries state that curricula shall be adapted to the pupil's level of development. Poland specifies this by saying that the content shall be adapted to the child's abilities of perception at a given age and shall guarantee the formation of socially acceptable attitudes. Bulgaria points to the fact that a child grows through a stage of sensitivity to the formation of ideas concerning objects, phenomena and processes to a superior level of synthesis. The Republic of Korea stresses the intellectual development of children and their ability to solve problems, and wants to foster scientific attitudes and satisfy intellectual curiosity. Czechoslovakia finds that the motivation and the level of ability of the pupil should form the basic elements in educational development. The Syrian Arab Republic wants to provide pupils with a minimum of scientific concepts and knowledge suited to their level of development, a scientific training enabling them to understand the environment, to find a scientific explanation of events taking place, to take part in environmental activities and to adapt to other disciplines in the curriculum. Viet Nam stresses that it is necessary for the child's level of development and the education given to correspond, which means that the knowledge provided shall fit the child's physical and psychological characteristics. Kuwait stresses the importance of an education adapted to the child's life, environment and degree of maturity. This creates interest and enables co-operation with teachers.

A science and technology curriculum adapted to the individual's environment is stated as a goal by Algeria, Cameroon, the Central African Republic, the Federal Republic of Germany, Jamaica, Malaysia, Mauritius, Morocco, Netherlands, Nigeria, Paraguay, Rwanda, San Marino, Seychelles, Tonga, Tunisia, Uganda, the United Republic of Tanzania, Zambia and other countries. San Marino states that such curricula shall be integrated into the social and environmental context of the child. The Federal Republic of Germany mentions that the objectives, content and methods of teaching are determined primarily by the needs and experiences of children in their environment, the requirements of individuals and social life, and the knowledge provided by science and technology. The German Democratic Republic stresses the link between education and real life, which ensures a preparation for creative life in society and develops the child's ability to contribute actively to social development. Uganda points to the importance of considering the child's future role in society and the expectations of the community.

That the curriculum must also be adapted to the individual's social background is emphasized by the Central African Republic, the German Democratic Republic, the Federal Republic of Germany, Jamaica, Malaysia, Mauritius, Nigeria, Paraguay, Rwanda, San Marino, the Seychelles, Tunisia, Uganda, the United Republic of Tanzania and other countries.

Very few countries have mentioned criteria which have to do with the *development of society* as such, and of science and technology in society.

However, adapting this education to changes in society and the world of work is stressed by Sweden. Mauritius emphasizes that it must be relevant to the level of

Renewal of science and technology education

development of science and technology in the country. Adaptation to changes in subject matter is mentioned by the Federal Republic of Germany, Paraguay, Poland and Sweden. The German Democratic Republic states that science and technology shall be developed on the principle of a scientific foundation, that is, its development shall be guided by scientific development, the findings of educational research and a uniform ideological approach.

The criteria stated for the development of science and technology education stress two fields: improvement in and modernization of content; and relevance to the individual child's situation. Very little is said about ideological, cultural and political training, or about societal development. Another surprising fact is that so very little is said about curriculum aspects like teaching methods and teacher training, or about other person-centred development goals like those of creating scientific understanding and skills, and of organizing better work situations in the schools. Most of these aspects have been mentioned by Member States in other connections, however, and countries may feel that enough has been attained and no need exists for stating such goals for the renewal effort. The re-phrasing of the way in which questions have been asked in the questionnaire from 'policies' and 'aims' to 'criteria' may be of some importance, but in all likelihood the effects are limited. This is shown by the fact that most countries use almost the same wording when talking about the aims of science and technology education as when discussing criteria for the development of such education.

Conclusions about goals and objectives in the renewal of science and technology education

Both the discussion in Chapter III on the goals and objectives of primary education in general and of science and technology within this education, and the present one on goals and objectives of the renewal process, show that the focus is on curricula and child development. The statements of goals and objectives of primary education showed a great deal of interest in all types of goals, but in respect of the renewal of education the goals and criteria stressed only a few fields.

In Chapter III, it was found that most educational goals concerned content, while the person-related goals had to do with the development of the individual child. Turning to the goals of renewal, we find that a further concentration with a clear focus on certain of the curriculum and child-development goals has taken place.

There might be several explanations for this, but the available material does not allow a specific explanation. A few words on the consequences of the situation are called for and the present authors will concentrate on science and technology education, which is the theme of this book.

More and more, science and technology have become essential parts of our culture, first in developed countries but later also in developing ones. Certain ideologies have incorporated the need for their proper use as essential elements, and there are also ideologies which take a cautious attitude towards certain results of scientific and technological development. Politically, the use of science and technology has become more and more important, since they confer both economic and military power. Under these circumstances, it is noticeable that no strong renewal efforts can be detected in this material with regard to their ideological, cultural and political applications. Questions pertaining to the introduction to science and technology education could include: in the future renewal of education, what role shall be given to science and technology in the description of culture?; what weight shall be given to encouragement of and warning against their use?; how shall these areas be incorporated into the ideological and political discussion in school?; and so on.

The responsible authorities answering the questionnaire were certainly aware of the fact that the renewal of science and technology education affects aspects of the life in schools other than the curricula and the pupils' situation, for example, the situation of other groups in the schools and of the school system itself. The need for the training of teachers and administrators has been mentioned, but much remains to be discussed: the new work situation of different categories of staff; the changed relationship between an often scientifically and technically illiterate environment and a scientifically and technically alert school; improved opportunities for schools to influence the development of the environment; and so on. It seems quite obvious that this development should be guided by means of an educational policy and an educational plan.

A third example of an area where statements of educational renewal goals and objectives are lacking is societal development. Education is a means to develop society — some find it is the only really important one — and science and technology education is particularly important for economic development. If the development of the country is to be planned, it is essential that plans — and accordingly goals and objectives — exist for the renewal of science and technology and education in the perspective of economic growth. There are many questions that need to be answered: what steps shall be taken to assure the availability of trained workers?; towards what fields shall the training in science and technology be directed to guide economic and social development?; is scientific and technological literacy less important than reading and writing literacy?; what role shall be played by the primary school (compared to secondary school) in such development?; and so forth.

MEANS FOR THE RENEWAL OF PRIMARY EDUCATION

The data reported above on goals and objectives for the renewal of primary education and on criteria for the renewal of science and technology education indicate clearly that nearly all countries aim for some kind of educational reform, which usually touches upon science and technology. Even if goals and objectives which will guide the development exist, there are still questions which must be answered before a complete picture emerges of the renewal effort. What is the scope of the renewal? Does it limit itself to changes in goals and objectives, or will it also involve curriculum content and the delivery process? Does the renewal process show a scientific and technological orientation? What steps are taken by countries to ensure the success of the renewal, for example, by involving different groups in it, by creating a research programme for studying the education system, or by teacher training?

The relationship between educational goals and objectives on one hand and educational outputs on the other is not direct but passes through curriculum content and the delivery system. This can be shown in a very simplified way in Figure 7. (See also Figure 5 in Chapter V.)

The translation of goals and objectives into curriculum content and structure is not necessarily easy, and there are several ways of doing it, sometimes with very different outcomes. Similarly, the creation of a delivery system with teaching methods, text-

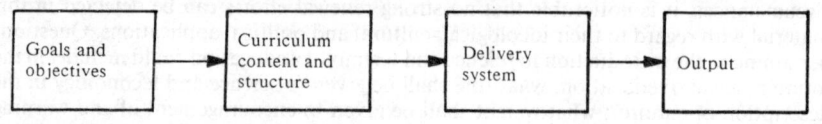

```
┌──────────┐    ┌──────────┐    ┌──────────┐    ┌──────────┐
│ Goals and│───▶│Curriculum│───▶│ Delivery │───▶│  Output  │
│objectives│    │content and│   │  system  │    │          │
│          │    │ structure │   │          │    │          │
└──────────┘    └──────────┘    └──────────┘    └──────────┘
```

FIGURE 7: Relationship between educational goals and objectives and educational outputs

books and materials, educational organization, teacher training and so on, opens up different possibilities which are suited to their purpose to different degrees. Even though considerable experience exists on how to do this, improvements in the system are often called for. One way in which educational authorites can check whether the system is efficient is by means of educational research. Such research studies the effects of different ways of translating goals and objectives into content and structure and into a delivery system. It can also evaluate the system by comparing outputs with goals and objectives

General scope of the renewal effort

Educational research has indicated that a renewal process restricted to one area, such as the introduction of a new teaching method, has usually very limited effects [91]. It has also been observed that a reform which at the outset deals with a specific aspect — like goals — often spreads and involves other aspects too. In the previous section of this chapter, an attempt was made to show the goals and objectives of the renewal effort as such. It can be seen that many countries describe innovations which would call for a renewal touching not only upon the goals and objectives of education, but upon curriculum content and structure, and on the delivery system as well. A separate section in the questionnaire used as a source for the present study dealt with the scope and content of the renewal.

In many cases the renewal has a very wide scope. Cuba reports an educational plan which takes into account the whole national system of education, with its different sub-systems, and starts from a consideration of the basis for the system: general polytechnical and work education. The renewal in Congo also affects all areas. In Switzerland, there are regional plans which cover goals, content and methods. In Nigeria, the renewal concerns goals and content reflecting trends towards technology, new methodology encouraging exploration and experiments, decentralization, improved school and community services, new teaching aids, new evaluation methods and so on. In Nicaragua, the change is based on the belief that the education system must be in line with the development of the revolutionary process, and the renewal involves objectives and principles, study plans, programmes, books and guides, buildings, materials, teachers and administrators and their training, and so on. In Colombia, the reform shall be in accordance with the necessities, possibilities and characteristics of each region. It will cover everything from objectives to teacher training and the distribution of textbooks and materials. On the other hand, Pakistan states that the main purpose of her renewal is to rectify what is now amiss, while not aiming for fundamental new goals or innovative content. The Republic of Korea also intends to improve conditions in the school so that the present situation — with overcrowded classes, large schools and shift schools — can be improved.

Elsewhere it is indicated that the *goals and objectives* of primary education shall be changed.

Paraguay mentions objectives corresponding to national reality, which incorporate advances within education and technology, and take into consideration the nature of the child. The primary school curriculum shall promote the creativity of the teacher, who is expected to adapt to the characteristics of the child and the area. Zambia stresses education that will develop the potential of each citizen for their own well-being and for the benefit of society and humanity. Malta mentions education of the child in acccordance with his/her abilities, and places stress on the training of handicapped children. The Bahamas mentions slow learners and deprived youth.

Belgium states five fundamental objectives, which may be summarized in the following way: (a) integrated basic education to develop the whole person by breaking down barriers between stages, classes and subjects; (b) individualization and differentiation adapted to each child's personality; (c) remedial education through individualization and remedial classes; (d) development of co-operation between school and family to compensate for retardation caused by the social milieu; and (e) stress on artistic creativity. In Egypt, the renewal will emphasize religious, national, moral and physical education and the relation between education and work. It will build on the environment, co-ordinate theory and practice, and so on. In the German Democratic Republic, the aim is to improve the quality of the curricula and to gear content to any new development in science and technology, and to reach the optimum personality development in each child. Training at the elementary level shall lay the foundation for further education. Czechoslovakia mentions the general goals of primary education: (a) defining the specific functions of the stages of primary education and of the aims, content, methods and forms of teaching; (b) enabling all pupils to pass through the system; (c) concentration on essential elements in education; (d) stressing the link between teaching, life and work; and (e) employing individual and group approaches.

A number of countries indicate that renewal stresses the goal of educating the whole personality of the child; the creation of equal chances for all children is also often mentioned. Madagascar, amongst other countries, stresses learning how to analyse and think in a critical way, and refer to the Malegachization of education. Jordan wants to promote group activities and increase vocational training which is dependent on group work. Rwanda aims for democratization with compulsory schooling and cultural development, including the use of the national language. The Syrian Arab Republic insists on national and humanistic behaviour among pupils and on the creation of a balance between their physical and intellectual development, and their acquisition of a general culture. Switzerland accentuates the harmonious total development of the personality. Subject teaching could be replaced with interdisciplinary teaching, while educational evaluation and selection criteria should suit the child, at the same time as strengthening co-operation between school and parents.

The renewal process in Cameroon aims for bilingualism, and the new goals also cover co-ordination between work and intellectual development, universalized education, and changed attitudes towards the reform and towards educational methods. In Malaysia, the goals include learning the 'three Rs', the development of attitudes and behaviour based on the human and spiritual values accepted in the country, the support of talent and leadership, and so on. Luxembourg stresses the integration of children of immigrants, the facilitation of vocational education and the furthering of enrolment in secondary technical education. The goal of the new democracy in the service of the people is mentioned by Senegal. In the United Arab Emirates, a goal is to engage pupils so that they participate actively in the educational process.

The *content* aspect of curriculum renewal is less thoroughly discussed. A number of countries mention specific subjects or areas, the most frequent being basic skills, the new maths and environmental sciences.

Among other types of answers: Mexico plans common content at the national level but also specific content for regions; Madagascar revises content in line with socialist ideas; the Seychelles uses the national tongue, integrates learning with work and adapts the content to national needs; the German Democratic Republic places more emphasis on moral and aesthetic education than before; Guyana stresses agricultural science and a scientific orientation; and Poland bases her renewal process on recent scientific developments and on the results of educational research. India mentions that her primary schools will have renewed curricula: socially useful and productive work will become an integral part of education; science syllabi are being upgraded; better use will be made of teaching materials and media; and scientific and technological development will be

Renewal of science and technology education

allowed to influence traditional practices and methods. In the USSR, increased attention is paid to science. In France, several subjects are involved, and the renewal mainly concerns the integrated subject *Activités d'éveil*, which aims to develop the child in many ways.

Quite a few countries have discussed the renewal of various aspects of *the delivery process*. Most answers refer to educational administration.

Although a great deal of the educational debate during the past few decades has dealt with teaching methodology and although much educational research is teaching-method oriented, few answers indicate fundamental renewal in this field. The Republic of Korea, however, has introduced educational technology. Poland and others mention the use of educational research data for the creation of books and materials and the choice of teaching method. The adaptation of the teaching/learning process to the maturity of the child, the use of physical and other activities, and self-learning are mentioned by Peru. In Iraq, there are workshops to enable children to become acquainted with practical work. Congo gives an ideological training to young people enabling them to accept adult responsibilities. Malaysia stresses self-development. Ethiopia wants to improve materials used in school and to consider leadership practices. In the answers, very little is said about innovations like individualization and group work, about the use of different media, or about the application of systems such as team teaching which may reduce costs.

There are not many references either to changes in educational structure. In a few cases the organization of the school system is being changed. Mexico has introduced a ten-year basic school and Egypt a nine-year one. Sri Lanka has restructured education to a 8+3+2 system and Nepal to a 5+2+3 system. Turkey is implementing an eight-year compulsory school in which moral and religious education is compulsory, and Algeria has initiated a nine-year compulsory school on a polytechnical basis. The primary school in the USSR is being reduced to three years but still within the previous compulsory system. Austria has changed its structure to whole-day schooling. Iraq and other countries have opened schools for retarded children. The Central African Republic has abolished streaming, and in the United Republic of Tanzania there has been a renewal involving management personnel and teachers. Earlier in this book, new ideas like the open school, new ways of combining subjects, ungraded school and so on, were discussed.

The most commonly mentioned change in management structure is decentralization. According to the plan in Mexico, this is seen as a means for improving the efficiency of primary education, strengthening community participation and reducing inequalities. Nicaragua mentions regionalization and nuclearization in the rural milieu, where out-of-school education shall be included and a move towards lifelong education is taking place. Japan mentions an increase in educational activities, to be provided at the discretion of the individual school, and a re-examination of curricula in the light of circumstances in each school and each community. Decentralization to make the comprehensive school more flexible and oriented towards practical applications is mentioned by Finland and Senegal. The Netherlands offers each school an opportunity to define itself in ideological and educational terms, outlining its principles and goals, and stating how these goals are to be achieved.

Renewal of out-of-school activities has been reported by a number of countries, and cultural activities with the family and community involvement by others.

Bulgaria draws particular attention to out-of-school activities within the framework of a social and political whole; they are organized in a creative way according to local and collective needs. In Poland, the educational plan takes into account co-operation with the mass media and cultural institutions. In Turkey, a literacy campaign, started in 1983, is tied to a primary education programme for adults.

Renewal with a scientific and technological orientation

Several countries stated that there is no scientific or technological orientation in the renewal process taking place. Another group of countries — Argentina, Chile, Cyprus, Japan, Madagascar, Peru, Poland, Rwanda, Sweden, Tunisia, the United Republic of Tanzania and Zambia — answered 'yes' without giving any further information about the nature and scope of this orientation. The USSR stated without specification that increased attention is being paid to science and technology throughout the system.

A number of countries report, however, that stress shall be placed on experimentation, skill training and practical work. This is the case with, for example, Finland, Spain and Uganda. Jamaica states that hands-on activity approaches are now generally accepted, and Nigeria has geared scientific orientation to the environment, giving it at the same time a broader base as the method of teaching has become more scientific following experimentation. Seychelles stresses the problem-oriented approach by means of which science is taught. In Iraq, laboratories have been set up in most primary schools. Australia states that the major direction in science teaching has been to develop process skills, that is, priorities are given to the study of scientific processes instead of products; hands-on experiences and discoveries are found important. A couple of Member States point to the introduction of scientifically and technologically suitable methods for teaching. Turkey states that study programmes, teaching methods and didactic materials have been continuously developed in accordance with scientific and technological principles, and that the methods used are based on and evaluated by means of scientific methods. The aims of science and technology education in Colombia include the development of a critical and analytical ability, with the child educated so as to be able to create, adapt and transfer the technology needed for the development of the country.

Practical work in close connection with science and technology education is emphasized by a few countries, for example, India, Jordan and the United Kingdom. The latter states that, according to policy documents, '...children should be given more opportunities for work which progressively develops their knowledge; and to introduce them to the skills and processes of science, including observation, experiment and prediction...'.

Nicaragua finds that two areas are of particular importance — natural sciences and agriculture — and projects have been launched to create materials, instruct teachers and contact communities. Paraguay hopes to bring balance between science and technology education and the humanities. Priorities are given to, among other areas, technical education and training for work. In Zambia there are projects where theories learned in the classroom are put into practice. These projects include out-of-class activities during which pupils make use of their talents. In Bangladesh, there are home-based activities involving parents and children, such as growing vegetables and fruit, and preparing food.

The relationship between the study of nature and the environment on the one hand, and science education on the other, is perceived by many countries.

In Bahamas, the new orientation of the science education programme includes biocentric environmental study stressing natural resources. Sri Lanka has increased the science content in the curriculum introducing environmental study to develop scientific attitudes among children. Belgium mentions science as a means for the exploration and understanding of the environment. In Bulgaria, children are taught the foundations of scientific knowledge and the laws of nature, and they are also brought into contact with nature and society, although scientific terminology is not used. When teaching general knowledge of nature, China uses the structure of the natural world as the framework; and in Senegal, all activities to improve the quality of education are based

some cases, the research effort is limited in scope. To a very large extent reference is also made to expert opinions, consultation with teachers and field experience.

Many countries indicate that their reforms are based on a series of preparatory studies. Spain reveals that the renewal programme was preceded by a period of consultation with and initial application by teachers with the aim of incorporating their observations into the final product. Guyana mentions classroom experiments and piloting in different subjects. Egypt reports a long series of research activities in areas like student performance and teaching methods, executed over more than twenty years. The prolongation of compulsory education was prepared by research. Innovations concerning special teaching in Malta were based on several years of experience in school. Austria reports that the present renewal is the fruit of extended experimental and research activities which have been going on since the early 1920s. The Federal Republic of Germany states that the primary education plan was prepared on the basis of a large number of expert opinions and studies by research groups and individuals. In Zambia, the renewal process was discussed in several fora, and committees were formed which studied education systems in several countries. In Thailand, the renewal of primary education administration was based on a study conducted by a governmental committee, and the renewal of the curriculum was based on studies by several agencies.

Czechoslovakia reports that research has led to a draft plan for development and renewal. The Byelorussian SSR states that researchers have been working for two years to improve science teaching and to analyse scientific principles on the upbringing and teaching of children, as well as methods of teaching certain subjects. In Turkey, the eight-year primary school was introduced after a pilot project of research and development work was completed. Other projects in the country deal with problems encountered with regard to compulsory schooling. Bulgaria points to two phases during the innovation process which call for research: the preparatory process, which needs research providing scientific arguments for renewal, and the execution process with research on methods, organization and other factors.

Cuba describes her renewal situation as a process preceded by diagnostic studies leading to a critical analysis of the educational situation and a descriptive model of the future school system. In Colombia, the curriculum revision is also based on a diagnosis of the educational situation. The recent decentralization in Argentina was preceded by fact-finding studies of administrative aspects and curricula. Mauritius reports that her renewal process has a research basis which concerns all national development projects. Basic research with testing in schools is carried out when new materials are devised. China has widespread experimentation with pilot studies leading to gradual extension, and local and regional studies are reported. The new ideas in the curriculum field in Mexico have as their starting-point opinions expressed by professors and educationists through a consultative process involving the formation of interdisciplinary groups.

Evaluation is also mentioned by several countries.

Belgium states that there is a need for evaluation by independent university teams, and Madagascar reports that her innovation process is based on periodic evaluation of the system. In Peru, the existing system is evaluated to determine its achievement level. There is an organization which, among other things, carries out evaluation and designs new plans and programmes for primary schools. Cuba reports an evaluation study on the results of the application of educational plans, which can then be adjusted.

Specific research problems have been mentioned in some answers: gifted pupils (Federal Republic of Germany); mathematics, science and language (Cyprus); immigrants, aborigines, handicapped pupils and open education (Australia); allocation of teaching hours (Finland); mother tongue (Seychelles); educational materials (Sri Lanka); teacher training (Syrian Arab Republic); and school management and the effects of using female teachers (Iraq). In Pakistan, there has been a project on village schools using an action-research model. Senegal has conducted major research projects on the introduction of a national language and science in school. In Malaysia, the research has mostly concerned the complexity of the present curriculum used in primary education. In Luxembourg, much research concerns post-primary education, language needs and migration. In the United Arab Republic, where research is a new activity, there are studies of inspection and student assessment. Many countries report research on science education.

The scope of the research effort varies a great deal.

The Central African Republic, India, Jamaica, Poland, the Republic of Korea, the United Republic of Tanzania and several other countries present a large number of research projects in fields relevant to the renewal of primary education. Australia presents a variety of applied and academic research carried out by universities and other bodies, but also uses data from abroad. Japan states that, with the co-operation of many educators, the Ministry of Education, Science and Culture has undertaken a wide range of studies of content and methods of teaching. The ministry has also accumulated data on school curricula by means of studies on their implementation. There is a national centre as well as many regional centres which conduct research and analyse data on curricula. Nepal reports several research studies that have been commissioned. In the Netherlands, subject experts are important when course content is determined, but there is a research programme dealing with educational and organizational renewal, including descriptive research, developmental studies and evaluation. France reports that the results of educational research play a large part in decision making.

Tunisia conducts experimentation which allows conclusions to be drawn about her new programme with its scientific orientation. In Paraguay, the Ministry of Education, with support from international organizations, has carried out several projects to raise the quality of education, improve enrolment and increase resources. In Chile, there is a centre which develops a research programme in support of educational renewal. To this are added the outcomes of work carried out by universities and colleges, and support from international organizations. Present projects concern, for example, decentralization and educational outcomes.

Teacher training for educational renewal

Any change in an education system depends on the teacher: preparation, personal qualities and attitudes; he/she constitutes a key element in the delivery process. Therefore, the training of teachers becomes a critical factor in the renewal process. Nearly all countries also indicate that teacher training programmes are adapted to changes in the primary school curriculum.

Many countries report that pre-service teacher training has been improved thanks to renewal in primary schools: the training courses take the latest trends into account, and try to respond to needs and to changes in curricula. A common statement about the teacher training programmes in Member States is that they introduce or strengthen science and technology education, place heavier stress than before on mathematics and introduce vocational subjects. A few countries report changes which deviate somewhat from this general trend, however, and the renewal taking place in teacher training varies a great deal between countries.

Argentina has decided to decentralize her teacher training to meet regional needs, and therefore a process of co-ordination between teacher training colleges and universities is planned. In Ethiopia, the experiment going on with a new primary school curriculum has been used as a basis for an intensive learning process in all groups concerned. Bulgaria uses two training methods for teachers of the renewed system: course themes to show future teachers essential points in the new system; and seminars to analyse educational documents dealing with the different subjects. In Czechoslovakia, young teachers take a one-year introductory course, and some established teachers take refresher courses in the interests of the renewal going on. In the Byelorussian SSR teachers take such refresher courses every five years, and self-education of teachers is important. Stress is placed on their knowledge of core subjects and their ability to master materials.

Nigeria employs a broad approach, and there is a new policy for the training of specialist teachers of science, technology and other subjects instead of generalists. A core programme for teacher training in integrated science has been developed. In the Netherlands, the reform involves an integration of nursery and primary education, and the training of nursery and primary school-teachers will be integrated.

Prolonged teacher training in general, or for some teachers, is reported by France, the German Demoractic Republic, the Federal Republic of Germany, Iraq, Jamaica and the Republic of Korea. The new three-year training programme in Jamaica involves both teachers and principals. In Iraq, there are new central teacher training institutions and new four-year colleges for lower secondary school graduates. Senegal has initiated a programme for pre-service training which considers the study of the environment and the initiation to science, technology and manual work in the new curriculum.

In Morocco, the Arabization process has brought about a renewal of teacher training. In Uganda, the renewal process often starts at the teacher training colleges, which means that the stage is set for pre-service and in-service training. The United States mentions widespread interest in a reform of teacher training as a result of research, and several colleges are working on a transformation of the existing research base into practical programmes for teacher training.

In a long-term perspective, improved pre-service training means an improvement in the quality of education and facilitates renewal, but in a short-term perspective, quality improvement and renewal must depend essentially on the professional development of teachers during their service. Most countries have therefore placed stress on in-service training and various ways of organizing such training are mentioned.

Seminars, briefing courses and study days are very common. Workshops are mentioned by Guyana, Nicaragua and Tonga. Special institutions are found in countries like Argentina, where numerous private training institutions help improve the quality of the teachers who return to them for further training. In the United Republic of Tanzania, teachers colleges are set aside for in-service training courses, and in the Netherlands, 50 per cent of in-service training facilities are reserved for teachers affected by the changes in primary education. In Zambia, there are teachers' centres organized by groups of schools using resource teachers for the dissemination of information. Guinea plans two new centres, and Iraq has three new centres with many facilities. In India, centres are set up which provide teachers with opportunities to consult materials and meet researchers, and to refresh their knowledge through weekend and vacation courses.

The use of radio and television programmes, the distribution of documents and other materials, and the publication of journals are mentioned by several countries. Core teachers are trained in the Republic of Korea to train others in their schools. In Gabon, there are seminars at the beginning of every year introducing methods to be tried out in the schools; teachers also follow periodic courses at in-service training institutes. China reports different methods to improve teaching: in-service study; a sabbatical term for training purposes; correspondence training; and training via radio and television. Algeria mentions short courses, study days and vacation courses.

Approaches to meet special needs have been reported: in-service training for the assessment of students in Nigeria; in-service training in remedial education in Malta. In San Marino, teachers are trained abroad, but the Ministry of Education organizes special courses to adapt them to the needs of local renewal.

Adequacy of the means

The presentation given above touches upon only some of the means available for the renewal of primary education. It is believed, however, that the emerging picture gives a fairly adequate description of on-going renewal. There are a few remarks which should be made about this renewal on the basis of the information received from Member States.

When the data on the goals and objectives of primary education renewal and the renewal of science and technology education were discussed, it was observed that they were concentrated on certain fields, while other goals like those of cultural, political, economic and social development, those referring to aspects of the school other than the child, and those concerning equality of educational opportunities, were rarely mentioned. Obviously, a school system can meet aims other than those expressed in documents, but this cannot be verified in the present study. The goals stated are rarely far-reaching but usually imply an extension of an already established trend. Exceptions exist, as has been seen, for example, in the *Escuela nueva* in Colombia, which opens its doors to people other than regularly attending pupils. However, in the area of science and technology education, there is little adaptation to the fact that the development in these fields changes our world and not only promises positive effects but also threatens our existence. In general, the data presented indicate that developments will continue within a rather narrowly determined frame.

Undoubtedly, curricula and teaching methods in many countries have undergone radical changes in the past few decades. Curriculum ideas, like subject integration and environment-related content, polytechnical education and practical work, have spread to many countries, particularly the developed ones. Methods like group work, individualization, team teaching, open-plan schools, the use of communication media, new types of textbooks and so on, have become important, at least in some areas. However, there is little material from Member States to indicate that fundamentally new ideas have been introduced; with few exceptions, the renewal efforts reported are built around well-known ideas. It also seems as if existing possibilities are rarely used. To take a few examples: team teaching, the use of programmed materials and other means for reducing educational costs are not often employed, even in poor countries; the use of ungraded schools to attack the problem of handling students of different ability levels is still only experimental; and modern media seem to be used mainly to supplement and illustrate what is said by the teacher, while attempts to use systems like loan video to introduce ideas which are unfamiliar to the teacher are not often reported. It is evident that there may be much that is going on in Member States which has not been referred to: new ideas that have not yet been tried out; provocative ideas that are too sensitive; or experiments tried out on so small a scale that they are not considered ready for presentation. About this we know nothing.

Much is definitely being done to present new course content, produce new materials and train teachers for new problems within the class, but many school systems still accept old concepts of education. When Sweden wanted to introduce technology into primary schools, this was felt to be an important innovation, but it took the form of a separate new subject. Many countries attempt to stress basic subjects, but these are still nearly always reading, writing and the use of the four simple rules of arithmetic — rarely science and technology. Educators all over the world worry about educational quality, but their solutions to this problem are often rigid examinations leading to repetition of grades and dropping out of school. There are many countries, mostly developed ones but certainly others as well, which have created efficiently functioning systems, yet in many developing countries the situation is serious. Twenty-five years ago, Coombs and his co-workers talked about a 'world education crisis'. Since then, education has expanded considerably, but the crisis has not been overcome. One reason might be that many education systems still show more faith in increasing resources from outside rather than changes in their own methods of work.

EFFECTS OF RENEWAL OF SCIENCE AND TECHNOLOGY EDUCATION

When innovations are introduced in an education system, there is likely to be a long series of consequences. If the renewal concerns educational goals, the latter must be translated first into curriculum content and structure, and then often into suitable delivery systems; if it is limited to the curriculum content or some other sector, in all probability it will cause changes in aspects which were not directly concerned. The more far-reaching the renewal, the more likely it seems that it will spread to other areas of the curriculum that were not originally involved. Furthermore, if the educational situation is changed, side-effects are likely to occur in student knowledge, working habits, skills, attitudes and so on. The changes are not necessarily confined to the educational area or subject for which the goals or the curricula were intended.

Curriculum changes
resulting from an introduction to science and technology

The existing curricula of science and technology education have been discussed in Chapters IV and V. They are always partly traditional and partly the result of recent innovations. In the questionnaire about the introduction to science and technology in primary school, the Member States of Unesco were asked to report what orientations and practical steps had been taken to ensure this introduction, for example, in respect of content, methodology and evaluation. The answers to this question gave information about curriculum changes that had been brought about by introducing science and technology. These answers are summarized below. A great deal of overlapping with what has already been said in Chapters IV and V is unavoidable, since in some countries the process has been going on for a long time.

The content of the primary school curriculum seems to have been changed in most countries.

The introduction to science and technology has caused Nigeria to create a core curriculum in science for the primary school system and a core curriculum in mathematics and science for teacher training. Mauritius has reformed objectives to fit the introduction of science and technology and changed the content of the mathematics, environmental studies and creative arts curricula. Poland reports the introduction of a new subject matter — 'socio-natural environment' — and the renewal of the curriculum for the subject 'work technology'. Bangladesh has a new primary school curriculum which includes science in the form of environmental studies. Bulgaria states that the initiation to science and technology, integrated in the new content, reflects progress in these domains. Algeria and Tunisia have revised their curricula with augmented time for science and technology, and Mexico has organized the content so that it can be strengthened and supplemented in higher grades.

In Czechoslovakia and the German Democratic Republic, as well as in other socialist countries, there are curricula which stress the socially useful character of manual work and training, and prepare the pupil for subsequent productive work. In the United Kingdom, the new orientation arises from the recognition that science and technology are important, with the emphasis placed on doing something practical and on the use of scientific investigations and experiments in school.

The need to adapt curricula to actual conditions has been stressed repeatedly in this book and is mentioned by many countries in connection with the introduction to science and technology.

In Uganda, content and methodology are regularly improved when needs appear, depending on experimentation and evaluation in the field. Jordan recognizes the contributions of specialists from the public and private sectors in the process of preparing the curricula for vocational education. China encourages experiments on science and technology education and mentions the use of extra-curricular activities.

The fact that science and technology are often integrated with other subjects is again stressed by a number of countries. Malaysia states that when the new curriculum was introduced, science lost its status as an independent subject, and its elements were included in language subjects. Even in the planned curriculum, it will be integrated with other disciplines. The effects of the introduction of science and technology on the whole curriculum is reported by a few countries, and Japan and Switzerland find that it is necessary to make a careful selection of essential content.

Few countries mention the use of new teaching methods in primary education. Activity-based methodology is stressed by Jamaica, Mauritius and Paraguay. Gabon and other countries structure the teaching/learning process around observations. Many other countries place the emphasis on skill learning and on activities like observation, measurement, classification, drawing conclusions, predictions and so on, as well as on communication.

The introduction of new content and new methodology in science and technology education has led to the introduction of new teaching materials, and a few countries mention this. Japan encourages teachers and pupils to develop teaching and learning materials. New science kits and equipment are mentioned by Nepal and Nigeria, teaching packages by the United Republic of Tanzania, and revised textbooks by several countries. Seychelles states that new materials have been prepared after extensive preparation stressing local application and the use of science and technology. Hungary produces handbooks on the background details of the curricula; organizes scientific conferences for teachers; offers educational materials, appliances, etc.; informs the public about the features of the new science and technology education; and establishes close co-operation with mass media producers. In Bulgaria, a national methodological centre has been founded, and the active collaboration of parents and authorities is assured. The Federal Republic of Germany stresses the importance of the mass media and points to the fact that children at an early age come into contact with technical devices.

A few answers mention new evaluation in the primary school.

In Japan it is specified that the results of teaching and learning should be continually evaluated, and pupils are to be evaluated in terms of the specific objectives of learning. Chile, Mexico, Paraguay and other countries report revision of and improvements to their examination systems, and Madagascar mentions inspection and visits to classes.

Effects in other areas

The introduction to science and technology in primary school cannot always be done without interfering — positively and negatively — with the life of the whole school. Therefore, it is reasonable to assume that there are appreciable effects in areas other than those directly concerned.

A few countries report that there are limited effects in other fields. Most of them have made small changes in their curricula, or have introduced the new subjects quite recently, or have a well-established tradition of teaching science and technology in primary school. On the other hand, far-reaching consequences are reported in other cases: effects on the development of the child; on the organizational structure of the school; and on textbooks, materials and equipment.

Improved language teaching is reported in many cases. The fact that science and technology teaching enriches the language, for example by adding scientific terms, is stressed. Ireland states that it promotes accuracy and rigour in the use of language and concepts and in the framing of questions. Rwanda finds that the learning of science and technology helps create a better understanding of concepts in the social sciences. Other countries indicate that it makes pupils more able to read, use books and dictionaries, etc., or that textbooks have been improved.

Improved scientific thinking is also reported for areas outside science and technology. Research carried out in the United States indicates this. The Republic of Korea mentions improved mastery of the investigative process, originality, scientific thought and attitudes, etc., and India stresses the cultivation of a scientific and rational outlook. Madagascar finds that, through science and technology education, the learner comprehends that all subjects are only instruments allowing him to master his environment. Japan states that such education helps children to develop positive attitudes towards examining problems with regard to natural phenomena, towards using their power of logical thinking, and towards trying to find the laws of nature behind natural phenomena. The Bahamas reports that pupils develop greater interest in and understanding of the 'what, how and why' in their environment. Abilities necessary for scientific inquiry, for example the ability to classify, measure, observe, communicate, formulate hypotheses, reflect and express criticism, draw logical conclusion and be objective, have been found to develop in children according to the Bahamas, Chile, Cyprus, Ireland, Jamaica, Mexico and Paraguay. Practical skills and knowledge of the environment are mentioned as effects by Egypt, as is an awareness of science and technology in daily life by the Bahamas.

Bulgaria refers to very important research findings in this respect. Young children manifested keen interest in science and technology and an increased ability to use knowledge from mathematics and science for a better understanding of technology. They also assimilated basic ideas about materials, instruments and machines in different technical, agricultural and service fields, thanks to the introduction of science and technology.

The realization that different subjects in the curriculum are related to and dependent on each other seems to have improved in a number of cases, particularly in countries with an integrated curriculum. Nigeria finds that the incorporation of science and technology into the primary school curriculum makes for a better understanding and an easy assimilation of the subject, and enables the pupil to discover the interrelationship of all subjects. Pakistan believes that the renewal of science and technology will have a direct impact on the methodology used in other subjects, with a shift from the dissemination of knowledge to the provision of relevant experiences. Switzerland finds that the impact of the new subjects on other subjects shows that all disciplines can be integrated without any of them suffering.

A very positive opinion about the effects of science and technology education on the child's personality development is expressed by Paraguay. It leads to: improved cooperation, responsibility and tolerance in individual and group activities; behaviour in accordance with norms, laws and rules; efficient use of time in study, work and recreation; and the practice of hygienic habits. The Federal Republic of Germany finds that the integration of different subjects into the *Sachunterricht* integrated subject often produces connections between scientific and technological matters on the one hand, and other subject matter on the other. Argentina finds that science and technology develop scientific curiosity in other subjects as well.

The effects on textbooks may be considerable. The Netherlands states that the content of textbooks used in language teaching is changing, and texts of language teaching and reading materials are being drastically altered. The effects on the teachers are mentioned by Peru: they can work more systematically, in a natural and practical way, as the technological focus enables them to understand the aims of the subjects, determine operational objectives and plan their teaching. Jordan reports that schools have been provided with workshops, and the United Republic of Tanzania explains that other subjects can be provided with teaching materials thanks to science and technology.

There are also organizational effects resulting from the introduction of science and technology. Argentina, Jordan, Turkey, Zambia and others report that the time devoted to other subjects has been affected, and Australia states that some teachers have indicated their reluctance to see an orientation towards science and technology dominate the curriculum.

Comments on the effects of renewal in science and technology education

The data collected on the effects of the introduction to science and technology do not allow very far-reaching conclusions. Few data on direct outputs are available. It is likely that there are considerable changes in student knowledge, habits, skills, attitudes and so on, as well as in other aspects of the school, but it is hard to see how to measure them. If goals as well as curriculum content and methods of teaching have been changed, the evaluation becomes difficult in the absence of stable criteria. (See Stufflebeam [78] and Werdelin [91].) The changes found in the curricula in different countries are hard to judge since it is not always clear what is a consequence of the renewal and what is part of the renewal itself. However, the above information indicates, as expected, that curriculum content and teaching methods have been changed as a consequence of the renewal taking place.

The information provided about effects in other areas is mostly based on personal judgement, often by people in administrative positions, even if some research results are reported. That the effects stretch beyond the fields of science and technology is not surprising. In the first place, these subjects take time away from other fields, which means that less time can be devoted to the latter; and secondly, the normal way of teaching science and technology in primary school in most countries is by integrating these disciplines with others. What is surprising is that so many countries agree that other subjects are being enriched by the renewal of science and technology education. It seems, judging from the reports, as if science and technology could be properly introduced or renewed in primary school without other subjects suffering, provided that proper preparations are made.

FRAMES FOR THE RENEWAL OF SCIENCE AND TECHNOLOGY EDUCATION

In Chapter V, the concept of 'frames' was mentioned: factors which influence an activity, whether positively or negatively. Negative frames are often called problems, difficulties or constraints, while positive ones may be called helping or facilitating factors. The questionnaire on the introduction to science and technology in primary education included two questions on the frames delimiting educational renewal. One asked about factors that hinder or accelerate the renewal process in general; the other dealt with problems and difficulties hindering the renewal of science and technology education.

There are many types of frames. Werdelin [89, p. 418] distinguishes between four major groups:

1. *External resources*: economic and material resources provided by people, authorities and organizations outside the system considered;
2. *Internal resources*: human and other resources available within the system considered, for example, the number, qualifications and experience of the teachers, and other categories of people in the system;
3. *Internal preparedness*: willingness and readiness of people in the system; here belong attitudes, values, opinions and factual knowledge of different people, as well as administrative and organizational factors; and
4. *External preparedness*: willingness and readiness of people, authorities and organizations outside the system; here belong laws, ordinances, agreements between groups, public opinion, political programmes and so on.

A few countries state that they have not encountered any difficulties in their primary education renewal. A further group of countries report that they have encountered few problems when introducing science and technology into their primary school curricula. This is the case with developed countries with a long history of educational renewal, for example, Austria, Finland, Japan, Luxembourg and Poland, but also Iraq, the Syrian Arab Republic and the United Arab Emirates. Most countries report difficulties both in respect of renewal in general and for the renewal of science and technology education in particular. A few countries state that their renewal efforts are accelerated by favourable circumstances.

External resources

For obvious reasons, developed as well as developing countries often mention a lack of resources needed for their educational renewal, while in other cases countries find that limited resources have prevented them from creating an optimal environment for this renewal. A very typical picture of the effects of being a developing country is painted by Pakistan. It is stated that the problem of renewal was originally underestimated. Further analysis showed that it was very complex and tied to the whole problem of development. The country has very limited implementation capacity with few professional people.

Lack of external resources can also be a problem when science and technology education are to be modernized. Limited resources, high costs, dearth of equipment and materials, lack of laboratories, scarcity of written materials reflecting the new curriculum, lack of buildings and rooms; these are a few of the factors which have been mentioned. In developing countries there seems to be a constant need for money and things that can be bought with money.

The provision of textbooks, guides, materials, equipment and facilities is inadequate in primary schools in many countries. Several states report a lack of scientific materials, laboratories and workshops. Sri Lanka mentions a lack of accommodation for pupils and inadequate transportation facilities, and Rwanda feels the need of new content and new books. Congo, India, Jordan, Qatar and Rwanda mention a lack of schools, and Zambia needs both classrooms and teacher accommodation.

India writes that during the 1950s and 1960s, many educators considered that a combination of programmed learning, teaching machines and audio-visual education would offer a significantly improved teaching system. This was not the case, however, because of inadequate equipment, inappropriate system design, lack of programming back-up and a real lack of operational potentials. In the Republic of Korea, there is a lack of all kinds of basic resources and teaching/learning materials which could improve science and technology education. Equipment for scientific and technological experiments and practical training are also lacking.

Cuba finds it necessary to make more unorthodox use of equipment and the installations that are found in schools. Algeria points to the slowness seen in the creation and use of infrastructures, and Paraguay mentions a lack of information sources for teachers and pupils, particularly in rural areas. A lack of written material with content in line with the new curriculum is mentioned by Chile, Cyprus and Sri Lanka; a lack of printing facilities by Uganda; a lack of curricular materials to support teachers in the classroom by Chile; a lack of educational infrastructure and buildings by Bangladesh, Gabon, Mexico and Uganda; a lack of laboratories by Chile, Jordan and Uganda; and so forth.

Viet Nam states that the country is still in a situation where it depends on small-scale agricultural production, which leads to shortages of different kinds. Bahrain points to difficulties caused by the distribution of equipment, the lack of infrastructure and an inadequate administrative structure. Pakistan refers to the ignorance and poverty syndrome, with people's energies being diverted to issues vital for survival.

Even developed countries sometimes mention a lack of external resources. The USSR states that there is some delay in the production of new equipment and teaching aids, and Czechoslovakia finds that the need for equipment is so great that it can only be satisfied gradually.

Internal resources

The quality and preparation of the teachers and administrators who are expected to execute the general renewal of primary education are often found to form bottlenecks.

There are countries which state that many teachers are either unqualified or inadequately prepared for the tasks imposed upon them by the renewal and they should be required to develop new skills and expertise in response to change. Some show negative attitudes when faced with working in a new situation. Paraguay finds it necessary to have regional teams which give permanent support to active teachers, as well as periodic courses, seminars and workshops which help in the implementation of the objectives. Sri Lanka reports constraints in the form of attitudes in school, while conservatism and misinterpretation of aims and objectives are also reported. Algeria, Morocco and other countries have also encountered resistance among teachers. Bangladesh mentions bad management of schools, and Mauritius states that the high mobility of teachers poses a problem.

France and the Netherlands mention the difficulty of creating participation of teachers in team work. In Belgium, teachers are sometimes hesitant about the changes that appear, and a slightly negative attitude may be due to a lack of relevant information.

Nearly all countries find that there are insufficient qualified teachers with adequate training in science and technology for the proper renewal of education in these subjects. Cuba states that it is necessary to improve the scientific and methodological training of teachers with the aim of achieving an optimal teaching/learning process, above all with regard to the scientific and technological character of education. Rwanda finds that, besides the inadequate number of generalists with diplomas in education, there are large numbers of unqualified teachers who can only make a difficult situation worse. This handicap, however, is found in other subjects and follows primarily from limited resources. The Republic of Korea mentions the over-burdening of science and technology teachers with duties. Bangladesh brings up inadequate administration, supervision and teaching.

Developed countries also report difficulties. The United States lacks science teachers with adequate preparation. Czechoslovakia states that the tendency towards speciali-

Renewal of science and technology education 173

zation is greater than that towards integration, both in teaching and in teacher training, which causes problems. In the Federal Republic of Germany, there is competition between different aspects of education, and the scientific orientation is in conflict with child-oriented teaching. The Netherlands states that the wide range of possibilities makes it difficult for teachers to make a choice and places heavy demands on the expertise of generalist teachers; in the country there is a diversity of teaching methods brought about by the freedom of the schools.

Internal preparedness

Organizational and administrative factors are sometimes reported as hindrances to the renewal efforts in primary school, as are attitudes.

In Chile, the lack of experience among local school authorities and school administrators is said to pose problems. Thailand reports a need to familiarize administrators with the nature and process of the new administrative structure. Nigeria mentions a lack of continuous support by research and data processing. In Senegal there is bad co-ordination with the research effort. An incompatability of organizational arrangements (class size, timetabling, space) is mentioned by Mauritius.

Bulgaria reports the problem of finding ways of using activity time in an optimal way. In the Netherlands, the renewal policy is constantly changing as experiences and new ideas are accumulated, and this sometimes gives an impression of inconsistency. Expert advice from supporting organizations is not always freely available. It is also declared that, within the framework of the given principles and goals, schools should create their own identity, which some find difficult. Spain discusses the lack of organizational flexibility in schools that should make the specializations and abilities of their teachers more widely available, and mentions the fact that technical directives are not prepared by the school principals.

Attitudes negative to the renewal are often found. Reports about this phenomenon are given by Mauritius and Nigeria, and Sri Lanka states as a problem the fact that parents do not find time to acquaint themselves with the changes. Senegal mentions old-fashioned attitudes in favour of traditional teaching. The renewal of science and technology education is frequently found to be hindered by a lack of preparedness on the part of the system itself. Algeria states that psychological and human frames have hindered the re-training of almost all educational personnel. Chile finds little interest in improving science education in the new school, particularly among generalist teachers. Egypt indicates that teachers have limited capacity to follow the constantly changing science programme, and Tonga reports a slow process of retraining teachers.

The United States reveals that low priority is given to science in primary school. The document from Bulgaria describes problems of a different kind: there is a conflict between the ability of the child to understand and the complexity of scientific matter. Many other answers point to the fact that it takes a long time and much effort to change opinions in different groups of people.

While these countries report difficulties, there are also descriptions of factors aiding the renewal process. In a few countries, for example Cameroon, Colombia and Turkey, educators and others realize that the education given is inadequate and something needs to be done about it. Guyana mentions the availability of resource persons and a general interest in renewal. Argentina points to the positive attitudes developed in educational communities when the effects of the renewal were felt. The United Kingdom stresses the fact that the reform is highly dependent on the endeavours of teachers and other groups.

External preparedness

External factors may hamper the renewal of the primary education system. Several countries refer to their difficult geographic situations, others to the characteristics of their populations or to the prevailing attitudes in different population groups.

Mexico mentions socio-economic factors in the population, and the United Kingdom refers to the decreasing numbers of pupils in her primary schools due to falling birth rates. France describes a situation where parents, trade unions, media and others divert the educational reform towards a development which lacks theoretical backing and is not always beneficial. Spain brings up the problem of politico-ideological resistance in the educational administration and in certain sectors of society, and Belgium refers to inconsistency in the administration of the renewal due to a lack of continuity in the government. Gabon experiences problems with the integration of the local culture into the new school, compared with its openness to science and technology. Two countries — Cyprus and Nicaragua — state that political conflicts with other countries cause grave problems, and Viet Nam still feels the adverse effects of colonialism and feudalism.

Several societal factors are reported to form obstacles to the renewal of science and technology education. Uganda mentions that a mainly illiterate population affects the application of science and technology, the Seychelles point to the low level of scientific literacy in the population, and Senegal reports a scientifically poor environment. Spain mentions an inadequate organization, and Belgium expresses a need to ameliorate the administrative structure. Malta brings up the lack of emphasis on primary science and technology education. Bahamas states that insufficient stress is placed on implementation in the schools, and Cyprus finds that the spirit of conservatism is present in any change movement.

Uganda points to the fact that research has so far failed to show that science and technology are essential to daily life, and several countries find that insufficient attention is being paid to science and technology in educational policy. The United Kingdom refers to the low priority awarded to it in the curriculum at a time when the government encourages the extension of science and technology education. France describes a delay of ten years in the issuing of instructions, which is the main explanation for schools having practically abandoned all that is not concrete subject matter and all that is not perceived as absolutely necessary. Ireland wants policy discussions on technologies and mentions computers as an example.

A number of countries report that external factors may help in the introduction of science and technology education. Australia believes that the autonomy given to the schools and regions is of importance since it facilitates appropriate responses to local issues. Malaysia finds it important that her renewal was initiated by the government, which can allocate funds. The Republic of Korea mentions the high expectations placed on renewal. Mauritius stressed the importance of the firm decision by the government to reform primary schools, the novelty of the content and methodology, and the flow of money from international agencies. The value of assistance from overseas organizations is also stressed by the Bahamas. In the United States, there has been a growing awareness of the performances of pupils and teachers, and a new emphasis on the fundamental aspects of education. There have been a large number of significant research reports.

Conclusions about factors influencing the renewal of science and technology education

Many frames have been reported to influence a country's possibilities of renewing its primary education in general and its science and technology education in particular.

Although no measures of the strength of these frames are given, it is known that they are often so strong that they can halt renewal.

Although even developed countries may report economic constraints, weak economies are characteristic of developing countries. Allied with a lack of resources are such effects as large classes, cheap systems like shift schools, inadequate teacher training, inadequate facilities, and lack of materials and equipment. All these phenomena influence the outcomes of the whole educational effort. However, since any renewal in science and technology education would call for costly preparations in the form of teacher training, new materials and better facilities, this education is likely to be particularly adversely affected. Many countries have called for international assistance to achieve the renewal they feel important [37]. Most of the resources must come from within the country, however, and since it is not likely that new resources will be made available, a redistribution of existing resources must be contemplated and means found of running the education system more cheaply.

The teacher remains a key factor in the renewal process. Data presented earlier in this book show that quite a few teachers in developing countries are inadequately prepared for their jobs. The situation is improving, but most of the teachers continue to have a humanistic background, and they are therefore badly trained for a role in the renewal of science and technology education. It is also obvious that many of them hold negative attitudes towards such a process, which is felt to hinder their work and make it more difficult. To be effective, in-service training, which is reported by most countries, is a long and costly activity and short courses often makes things worse. Means for the improvement of the teaching situation for inadequately trained teachers and for the improvement of their qualifications must be sought.

Administrative and organizational factors have also been found to be of great importance. Administrative changes are being undertaken, for example decentralization and regionalization, and they might prove effective once established. Organizational solutions like multi-grade classes, open schools or classrooms, team teaching and non-graded schools are described. They might permit good teaching to take place even in poor areas and they could also offer teachers the possibility of using their special qualifications to the best advantage.

The readiness of society for educational changes is sometimes questioned. Education is frequently a politically sensitive area; teachers' organizations are often conservative and defend the interests of their members with vigour; and parents and others are often reluctant to accept educational renewal which deviates from what is felt to be the 'right education'. To these have to be added difficulties caused by the climate and the geography. In spite of these problems, many countries seem able to carry through important renewal projects; some even report that little resistance was encountered.

EDUCATIONAL RENEWAL AND EDUCATIONAL PROBLEMS: A SUMMING UP

In Chapter II it was shown that the education systems in many, perhaps most, countries in the world experience grave problems. It was shown that in many parts of the world large numbers of children never get any education, that many of those who enter school drop out after only a few years, that repetition of grades is common, that educational inequities exist, and that the achievement level of the pupils is often low.

This chapter has dealt with the renewal of science and technology education in primary school. It has touched upon its goals and upon the means used in the renewal, and an attempt has been made to illustrate some of the effects of the latter and to estimate to what extent it is surrounded by frames of different kinds.

The introduction of science and technology education — or the improvement of such education — cannot solve all the problems mentioned; in this respect quite different efforts are necessary. It *can* contribute to solving some of the problems, however. It has been shown that the goals of renewal are often concentrated upon producing improved curricula and creating the conditions for desirable child development. Very little is said about other aims of education, but certainly these are already important.

However, the authors believe that educators and educational politicians see science and technology education in a wider perspective. One of the main reasons for the malfunctioning of an education system is a lack of resources — external ones like money and internal ones like qualified teachers. But the education system contributes to the improvement of the country particularly through training people in applied fields like science, technology and practical work.

Another reason for malfunctioning is that the education system is often perceived as an alien body in society since it provides schooling which is impractical, theoretical and even dangerous to cultural values. One way of making the schools more relevant to societies would be to ensure that what is taught is applicable to the immediate environment, and training in science, technology and practical work should play a key role.

The present renewal is reported to have characteristics which are in line with these thoughts: science and technology education is becoming practical and environment related; emphasis is being placed on practical work; decentralization and adaptation to local conditions are taking place; and integration with other subjects enables a global approach. It has been remarked earlier in the text above that science and technology are still treated as orphans in the family of disciplines in many countries. It has also been found that the renewal of science and technology is limited in scope and often follows rather traditional routines. It can be added that until recently very little was known about the results of the renewal efforts. The limited scope and lack of daring in the renewal are certainly due partly to the fact that modern science and technology education are recent phenomena in many countries, and partly to the frames which have been discussed above. The former will change, and the latter must be changed if countries are to develop properly.

CHAPTER VII
Reflections and questions for deliberation

The thirty-ninth session of the International Conference on Education, which was held in October 1984 in Geneva, provided delegates from 118 Member States and myriad other organizations, both governmental and non-governmental, with an opportunity to discuss some of the most pressing problems and opportunities confronting the world of education today.

It is clearly impossible to convey on paper a complete analysis of that discussion, although the *Final report* of the Conference represents a valiant attempt to do so [34]. However, each delegate and observer has his or her own thoughts and interpretations of those discussions. And, following a period of reflection, both on the conference and on the data set out in the foregoing chapters, we intend in this final chapter to share our personal interpretations and reflections. The purpose is not, of course, to draw out 'implications for practice' such as are found at the conclusion of some research studies. Given the disparate nature of educational policy and practice around the world, developing generalized recommendations for all countries would be both pointless and presumptuous. Rather, in the following pages, we intend to use a variety of theoretical perspectives from the educational literature to raise questions that others may find useful as they examine their own educational situations. (For more on theoretical perspectives and their use in educational research, see Roberts [65] and Munby [48].)

For the purposes of clarity, we follow the same sequence of topics as presented in Chapters II to VI.

CONTEXTS OF SCIENCE AND TECHNOLOGY EDUCATION IN THE PRIMARY SCHOOL

Science and technology are taught in the broader context of education systems which in turn operate within national and international environments. These environments are characterized by geographical, economic, cultural and many other factors that together constrain the ability of any country to achieve its national objectives for education. Nevertheless, policies are in place in nearly all countries which declare universalization of primary education to be a national goal.

Furthermore, the data presented in Chapter II show that progress toward this goal is being made. Enrolments (see Table 3) are generally increasing and especially so in parts of the world where they have traditionally been very low. Over the twenty-two years shown in the table, primary school enrolment rates in Africa have increased by 83 per cent. In another comparison, the gap between enrolment rates in developed countries and those in developing countries was reduced from 33.7 per cent in 1960 to 15.5 per cent in 1982. This progress is being supported in all countries both through policies for compulsory free education at the primary school level and through funding to support such policies. The percentage of gross national product (GNP) being spent by Member States on education has steadily increased during the past twenty-two years — and more so in developing countries than in the developed countries (see Table 7, Chapter II).

Yet, problems persist. The recommendation adopted by the Conference warns that the universalization of primary education 'should not lead to a lowering of standards or of the quality of education' nor should progress 'be considered in quantitative terms only' [34, p. 28]. These concerns are well articulated by V.G. Kulkarni of Bombay in a recently published paper. He writes critically of some attempts to achieve the goal of universalization:

Having realized the importance of universalizing education, an all out effort is launched to achieve this objective. However, the perception of the task is basically wrong. It is conceived as a programme of giving more of the same to many more people. The emphasis is, therefore, placed on generating infrastructures, solving problems of management and, above all, on creating and mobilising resources. Aspects like undertaking research to identify the real stumbling blocks, increasing the relevance of learning tasks to real-life situations, and relating education to productivity are often sidetracked. In other words, dissemination of a traditional package to cover larger sections of population gets precedence over identifying the contents of a package that can be universalized profitably. It is the contention of this thesis that ignoring these aspects has been one of the major causes for the alarming drop-out rate, wastage and stagnation in education, in spite of a tremendous effort to generate a gigantic infrastructure [40].

The data provided in Chapter II on wastage in education provide somewhat discouraging confirmation of this thesis (see Tables 5 and 6).

How, we may well ask, can universalization and renewal of primary education be conceptualized in qualitative terms when even the word 'universalization' seems to invite a merely quantitative analysis. In our judgement, it requires that we step back from this immediate question and clarify what we mean when we talk of education, the quality of education and universal education.

First, we should recognize the tendency, alluded to in the opening chapter, for us to use the words 'education' and 'schooling' interchangeably. The concept of education becomes narrowed to what can be achieved in schools, or (worse) to what *is usually* achieved in schools. And the institutional constraints of schools become limits to what is regarded as education. In our review in Chapter II of the worldwide influence of the European model of schooling we noted five features: fixed entry age; full-time; teacher dominated; one grade per year; examinations. It is interesting to note that, whether or not any of these are essential or intrinsic to education, all of them are characteristic of schools the world over. Furthermore, these features lend themselves well to quantitative measurement (enrolments, drop-out rates, time spent in school and so on). For the most part, these measures are simply indicators of the operation of a system of schooling. This is not to say, of course, that we should not watch enrolment trends or test student achievement. It simply warns us against taking limited measures of the operations of a school system as true indicators of the quality of education being received by our children.

Education as initiation

R.S. Peters, the educational philosopher, has argued that it is inappropriate to think of education simply as the means to some end beyond itself. He claims that education is not essentially an instrument for a society to solve its economic or social problems but an enterprise whose aims can be found within its own activities. He goes on to propose that education be thought of as *'initiation* into worth-while activities and modes of conduct' [57, p. 34]. We need not delve further into the complexities of this concept that Peters develops. Consider, instead, some of the straightforward implications of accepting this concept as the essence of education.

First, the notion of initiation suggests some kind of progressive process that is not limited to a few designated school years but can last for a lifetime. It also suggests a process that can take place in informal settings as well as in the formal context of schools. This view of education is consistent with the recommendation of the Conference regarding the complementarity of formal and non-formal education [34, p. 28]. The concept of education as initiation also contains echoes of our discussion of technology and technology education in Chapter I and implies a sort of intellectual, moral and physical apprenticeship.

The second implication of accepting this concept of initiation is that it focuses attention on the substance of education and not merely on its external features which, as we have noted, are simply indicators of the operation of school systems. Initiation invites the question: 'Initiation into what?'. While the rest of this chapter extends this discussion in the areas of science and technology education, the Conference debates and recommendation give clear indications of the intrinsically valuable activities into which Member States wish their young people to be initiated. These include, but are not limited to: *reading* — the worldwide focus on literacy is evidence of the realization that individuals who are not initiated into this activity are cut off from access to many others; *reasoning* — the process by which one learns to handle words and ideas logically and persuasively; *work* — the activity through which individuals make a livelihood for themselves and a productive and socially useful contribution to their society; *learning* — the way to the development of new knowledge and skills throughout a lifetime. Each of these can be seen as intrinsically valuable, as good in itself and not merely as the means to some remote end. Furthermore each is an activity one is initiated into through an apprenticeship process.

Finally, the concept of education as initiation suggests some new criteria for assessing the performance of schools. It is characteristic of the apprenticeship process that one learns from it by attaching oneself to a 'master', by observing his actions, by copying him and by submitting to his correction. Over time, one learns what he knows and becomes able to do what he can do. He is the key to the process. However, the only part of formal schooling that frequently operates on this model is the doctoral programme of a graduate student. For the most part, formal schooling, in its laudable attempt to become universalized, has lost this sense of a master demonstrating these activities to his students. The modern-day teacher is more like a signpost telling his students where to go than a guide who says, 'Come with me and we will go together.' If education is truly to become initiation, as we believe Member States have indicated in the Conference recommendation, then we must collectively rediscover the teaching role of the apprentice master. Reasoning will then be 'taught' by individuals who are truly rational, reading by those who love to read, attitudes to work by those whose work is productive and exciting, and learning by those who have never stopped learning themselves.

Thus the true indicators of the quality of education are to be found not by simply counting heads, nor even by scoring examination papers, but by observing the educational process itself and asking about its 'authenticity', as one group of educational scholars has termed it [49]. If these more stringent criteria are used to examine schools, then even the wealthiest nations may find cause for concern. (See, for example, the case studies of science education in the United States [76] and in Canada [51].)

Questions for deliberation

On the basis of the data set out in Chapter II and the above analytical commentary, we suggest the following questions for educators and others to ask themselves in regard to their own school systems.

1. Into what activities, attitudes or modes of conduct do we expect primary education to initiate our children?

This is not quite the same as asking for a statement of aims and objectives for primary education. It implies, rather, that the schools are there to provide intrinsically valuable activities for children and that it is useful to identify what these are and why they are considered to be of value in any given cultural or national context. These activities can include such generic skills as reading and reasoning, forms of knowledge such as science and history, as well as attitudes such as patriotism, an appreciation of literature and so on. This clarification is an essential prerequisite to any qualitative analysis of the performance of schools.

2. Are there activities, attitudes, etc., into which we want our young people to be initiated but which are best carried out by agencies other than schools?

One of the themes to which we shall return later in this chapter is the limits of what can be achieved by schooling. We have already established a distinction between the more generic concept of education and the more limited one of schooling. Not only is it important that we ensure that the schools are equipped to achieve what we expect of them, it is also important that we do not expect the impossible. For example, we might agree that developing appropriate attitudes towards work is a desirable component of primary education. It would not be inconsistent with that conclusion to determine that industrial, agricultural or other settings outside the school were the best places in which to initiate young people into such attitudes. In other words, schools will be recognized as places where *part* of a child's education takes place, other parts being carried out in the home, at work and so on. This view is often acknowledged in principle while, in practice, schools are operated in isolation from the other educative opportunities a child experiences.

3. How is the term 'primary' understood when applied to education as distinct from its application to schooling?

Primary schools are, as Chapter II demonstrates, variously defined by Member States in terms of ages of children at admission, numbers of grades or years, and so on. This question concerns not the institutional characteristic but the educational concept of primary education. Traditionally, education has been something an individual 'received' during the early part of his or her life *after which* the rest of life — career, family, etc. — followed. More recently the concept of lifelong education has become popular in recognition of the benefits to an individual and to society of continuous initiation into new skills, new attitudes and new activities. Yet, to what extent have the policies and practices of primary *school* reflected this concept? Is 'primary' still taken to mean 'before secondary' or does it actually reflect the idea that each individual is being initiated into a lifelong educational process?

* * *

These first three questions are designed to address our basic concept of primary education and to sort out some of the differences between education and schooling in any given national context. However, most of Chapter II is taken up with a discussion of indicators of the performance of primary education. It is appropriate therefore that we conclude with some consideration of how we can assess the quality of educational performance in our schools.

4. *Given decisions concerning desirable educational activities, attitudes, etc. (see Question 1 above), what observations or measurements could be made that would serve as indicators of the quality of primary education?*

One of the indicators discussed in Chapter II was wastage through dropping out and repetition. Kulkarni's thesis, set out earlier in this chapter, was that such wastage was due, in part, to the universalization of inappropriate forms of education. Therefore, we may ask:

5. *What empirical evidence exists concerning the reasons for dropping out, repetition, etc.? Can educational research help to substantiate or refute the Kulkarni thesis?*

Another problem of primary education identified by the Conference and discussed in Chapter II, was discrimination on the basis of sex, race, language, physical handicap, geographic location and other factors. Few countries acknowledge explicit and intentional discrimination against specific groups in society. However, many recognize that implicit, non-intentional discrimination of many varieties can occur.

6. *What groups are being discriminated against in education and what steps can be taken to reduce this?*

The penultimate question concerns the means by which all of these important educational policy issues are resolved within society. While every country has its designated authority for *deciding* policies in education, there are many interested parties or stakeholders in such policy decisions. Often, however, those who must decide are not well-informed about these various interests. Thus, while a decision can be taken, its implementation may be difficult because of entrenched attitudes and practices. This is recognized in the Conference recommendation that calls for 'effective participation by all sectors of society in the process of universalization of primary education' [34, p. 28]. We therefore conclude by asking:

7. *What steps can be taken to ensure maximum participation in the deliberative process leading toward universalization and renewal of primary education?*

GOALS AND OBJECTIVES OF PRIMARY EDUCATION

The concept of education as initiation into essentially worthwhile activities is not inconsistent with the principle that educators should pursue goals or objectives. Peters insists that the true worth of education is to be found in its own intrinsically valuable activities rather than in its being a means to some external ends (such as the improvement of society). This is important when one seeks to justify the content of education. However, educators need to have goals and objectives in mind, if they are to operate in a purposeful manner and to make choices among alternative teaching strategies. And while the first section of this chapter focused on a general approach to the justification of educational activities, this section deals with what it means to have goals and objectives in mind when planning a curriculum or teaching a lesson.

Chapter III reviewed the goals and objectives of primary education as these were reported by Member States, the place given to science and technology within those goals and objectives, and the objectives of science and technology education itself. The result of this review was such a complex mass of information that a brief summary is, perhaps, appropriate here.

The number and variety of goals for primary education were so large that a relatively complex model or category system was required to classify them (see Figure 3, Chapter III). One dimension recognized the range of specificity of goals and objectives (from relatively general to very specific). The other two dimensions were of more substantive significance:

Person-centred ↔ Society-centred
Tradition-oriented ↔ Change-oriented

The first of these distinctions differentiates between goals that were focused primarily on the needs of the individual and those that derived from the needs of society as a whole. The second dimension distinguished between goals having to do with the preservation of something and those oriented towards change or development. Each of these bi-polar dimensions can be combined to generate four categories of goals as follows:
1. Society-centred/tradition-oriented — ideological goals.
2. Person-centred/tradition-oriented — educational goals.
3. Person-centred/change-oriented — personal change goals.
4. Society-centred/change-oriented — societal change goals.

Within each of these categories, examples of general goals and more specific objectives for primary education were provided from the responses of Member States to the IBE pre-Conference questionnaire.

The second component of Chapter III was an examination of these statements of goals and objectives of primary education to determine the place, if any, of science and technology. The rhetoric of modern education contains many references to the scientific and technological world in which we live and to the need to help children prepare for life within that world.

Furthermore, we have argued (in Chapter I) that the key to economic and social development lies in the partnership of science, technology and education. Given this background we expected that governments might stress science and technology in the goals and objectives of primary education but, in general, as Chapter III explains, we found this not to be the case. One might conceive, for instance, of each of the four categories of goals identified above embracing a science and technology component. Yet, this is rarely articulated in policy statements.

The third part of Chapter III was devoted to the (separately reported) goals and objectives of science and technology education. It is perhaps not surprising that, once a particular subject-area of the curriculum is identified, most of the objectives fall within the category we have called educational. Yet examples of objective for science and technology education were identified in each of the other three categories also. This finding parallels the conclusion of a similar analysis of goals and objectives of science and technology education in the ten provinces of Canada [54]. While the rich detail of the international survey was obviously missing in the Canadian study, the same diversity of objectives was evident.

It is clear that much is expected of primary education in general and almost as much from the science and technology component of that education. Are these expectations realistic or do we face inevitable disappointment and frustration? Let us move towards answers to these questions by considering three dilemmas that emerge from our account of the goals and objectives of primary education.

THE DILEMMA OF DIVERSITY: THE SCHOOL'S DILEMMA

If practicing educators were to review the list of educational goals presented in Chapter III, they could be forgiven for wondering just what sort of miracle-workers their societies and governments take them to be. The first reason for such thoughts is that, as practicing educators, they assume that the list of goals for education is the same as goals for schools. And they know that the schools in which they work could never achieve what is represented in that list. That is not to say that the educational goals articulated by our governments (on behalf of the societies they serve) are not desirable goals. It is simply to underline the distinction, referred to earlier, between education and schooling and to point again to the necessity of determining consciously how all of our educational goals can be achieved.

Professional educators have lived with this problem for a long time. Societies' expectations of schools have been escalating throughout this century as any curriculum historian can show. And, in our own day, the troubled economic circumstances in which many countries find themselves are likely to raise the expectations of schools yet again. The way in which the profession has dealt with the problem of unrealistic goals is through a system of progressive filters. These were cited in Chapter I from the experience of the Canadian study and are summarized again here: *intended curriculum* — as set by ministries of education; *planned curriculum* — as hoped for by teachers and textbooks; *taught curriculum* — as actually delivered in the classroom; *learned curriculum* — as achieved by the students.

It was the experience of the Canadian study [72] that the ambitious expectations of government, which comprise the intended curriculum, are steadily reduced through the planning processes of curriculum development (which tends to ignore many of the goals), through the constraints of the classroom and through the many factors that affect what students actually learn.

Three brief quotations from the case studies that formed part of the Canadian research programme illustrate well this reality of professional life.

With only one exception, provincial curriculum guidelines were conspicuous by their absence in all discussions with the Prairie High science teachers. Teachers have adapted these guides to suit their own practices [1, p. 258].

Mr. Blake would like to have the students actively involved in investigations on a more regular basis.... He explains his dilemma: 'I would like to approach science as being an activity, but I'm not always able to do it.... I have to make compromises.' [69, p. 54].

I suspect that [teachers] are [not] aware of [the] potential to damage their students' cognitive development by producing blind submission to authority, dependence on experts, abdication of inquisitive thought, absence of genuine criticisms, over-estimation of rationality (other people's, to boot!), feelings of helplessness and so forth [82, p. 244].

The first of these illustrates how the intended curriculum can be mentally omitted when concrete planning takes place; the second, how even the teacher's planned curriculum has to be modified in the classroom context; the third quotation concerns some of the possible negative by-products of science teaching.

This account of the anatomy of the difference between education and schooling is not to justify the attitudes represented by the teachers cited here, nor to argue that schools cannot achieve more of the educational goals that are set for them. Rather, it is to suggest that, in identifying how all those desirable educational goals can be achieved, realism is important.

Questions for deliberation

8. Which of our educational goals are appropriate for schools, for primary schools, for science and technology programmes within primary schools?
9. Which agencies other than schools can assist in the education of our children, in their primary education, in their science and technology education?
10. What steps can be taken to reduce the gap between intended curriculum, planned curriculum, taught curriculum and learned curriculum?

The science policy/education policy dilemma

This issue can be stated with ease but, we suspect, resolved only with considerable difficulty. The external indication that there might be a problem lies in a curious combination of facts: (a) governments state that science and technology are of great importance to the economic and social well-being of their people; but, (b) the general goals of primary education often make little mention of science and technology; and (c) the more specific objectives of science and technology education frequently have little to do with the broader economic and social context.

The data presented in Chapter III make it clear that this problem is not found uniformly around the world. The socialist countries, for example, incorporate a polytechnical principle in their systems of primary education and make frequent reference to the importance of linking education with productive and socially useful work. While some other States oppose the inclusion of such references within international recommendations, it is also true that, within these very countries, concerns are frequently expressed by industry at the lack of connection between education and the world of work.

Why does the world of education, including the world of science and technology education, have such difficulty in meeting the world of economics and industrial development, of which science and technology now form an increasingly important part? We think that many countries face a problem that has to do with the structure of government or, more strictly, of government bureaucracy which frequently divides policy making into separate areas corresponding to a distinction between economic and social policies. Given this distinction, science and technology policy is usually seen as a component of economic policy, whereas educational policy (including policies relating to the teaching of science and technology) are considered to be social policies. The problems occur, of course, in the gap between the two, in which little policy making takes place. Hence the need for special commissions, privately sponsored forums or other means to cross this 'great divide'.

In terms of practical policy, the consequence of this division is that policies for science and technology *education*, having vitally important implications for the economic development of a nation, may be left to individuals for whom science is only one subject in the school curriculum.

This is not only a problem for industrialized countries. King has recently argued that the same division underlies major contradictions in the development of coherent polices for science and technology education in developing countries [38]. We therefore ask:

11. How does the structure of policy making in government affect the development of policies for science and technology education?
12. What steps can be taken to ensure that any structural gaps in policy making for science and technology education can be overcome?

13. *To what extent does the existence of separate divisions for education and for science and technology within international organizations such as Unesco tend to perpetuate any weaknesses in 'interdisciplinary' dialogue within Member States?*

THE DILEMMA OF DIVERSITY: THE TEACHER'S DILEMMA

If the diversity of general educational goals and objectives puts the schools in an awkward position politically, it also places the classroom teacher in a professional impasse. As we have seen, teachers have their way of quietly ignoring the more unrealistic expectations that governments place upon them. They filter the ideals until what they regard as realistic remains. But *can* science and technology be taught with a variety of different educational objectives in mind or is 'learning the content' the only objective that school science and technology can achieve?

This question has been the focus of much research in recent years by Roberts based on the concept of 'curriculum emphases' [64]. Briefly, the concept is that science teaching, like most other forms of communication, conveys more than one set of messages to its recipients simultaneously. At one level, science teaching obviously communicates scientific information. But, beyond this level of communication, science content is always embedded in a contextual web of intent or purpose. The various contexts in which science topics can be presented can result in correspondingly different learnings on the part of the student. These different 'contexts' have been described by Roberts as 'curriculum emphases'. He explains the term as follows:

A curriculum emphasis in science education is a coherent set of messages to the student about science (rather than within science). Such messages constitute objectives which go beyond learning the facts, principles, laws, and theories of the subject matter itself — objectives which provide answers to the student question: 'Why am I learning this' [64, p. 245].

And while part of any science teacher's concern is for teaching the content of science to students, another — frequently an even more important part — is that the content shall be learned for some purpose beyond itself.

The example of separate treatments of the same science topic in two different textbooks illustrates the point. In one textbook, all of the information about the methods of heat transfer (conduction, convection and radiation) is set in a context of explanations about how refrigerators and solar heaters work. Photographs and 'cut-away' diagrams illustrate these explanations. In the other, the same scientific principles are discussed but, this time, accompanied by a historical comparison of different (and competing) theories to explain thermal phenomena. This account is illustrated with excerpts from eighteenth century scientific reports. In each case, the science content being communicated is the same while the contextual communication — the curriculum emphasis — is markedly different. And these different emphases are the concrete representations of each book's different purposes when a student is learning about science.

The concept of curriculum emphases has been shown to be useful for classifying the goals of science education and clarifying the concept of scientific literacy [66], as well as for interpreting different ways of teaching science in the classroom [67]. In the present context, it is useful as the link between the goals and objectives of science and technology (discussed earlier) and different approaches to the teaching of science and technology (Chapter V and below). While a diversity of goals and objectives may appear daunting, their achievement is possible if an equally broad array of teaching strategies are used. If, however, teachers have been trained to think in terms of one 'right' way to teach science, then it is probable that only one curriculum emphasis will be presented to students and only one set of goals reached. The curriculum emphasis concept can be a means, therefore, for bringing the 'taught curriculum' closer to the 'intended curriculum'.

Questions for deliberation

14. How many different curriculum emphases can reasonably be incorporated into the planned curriculum of science and technology education at the primary level?
15. What is an appropriate balance among curriculum emphases (goals) for science and technology education at the primary level?
16. Do teachers think of only one or two curriculum emphases as appropriate or do they recognize and incorporate a range of emphases within their teaching?

STRUCTURE AND CONTENT OF PRIMARY SCHOOL CURRICULA

The vision of science and technology education emerging from Chapter III is one where both subjects function as vehicles for reaching the goals of primary education enunciated earlier. As was described, the concept of curriculum emphases was designed to help educators conceptualize how this might be done. The concept stresses that there should not be a choice between learning the subject, in this case science (or technology), and reaching certain objectives beyond the subject. Both of these desirable goals can be achieved simultaneously.

However, the picture in Chapter IV is one of primary education containing little science and less technology, when the presence of these subjects is assessed in terms of time. Furthermore, the strong impression from the data is that, where science is taught, it is taught largely in order that students may acquire scientific *information*. This reality falls far short of the intentions and plans for education *through* science in primary school.

Technology suffers, if anything, a worse fate. It is taught with an almost exclusive focus on manual skills, leaving far behind all the rhetoric concerning the technological world in which we live. In some cases, even the manual skills for which training is provided are obsolete in the societies where they are taught.

How, then, can curricula for science and technology be designed so that the educational goals for primary education can be achieved using science and technology content as the vehicle? This is the problem for which the concept of curriculum emphases was designed. Accordingly, the following section develops the discussion of this concept with a series of three examples of different emphases for each of science and technology. Each embodies its own criteria for the selection and ordering of content. Following the discussion of the different emphases, we shall examine some of the criticisms which have been directed toward the present day science and technology curricula. The section concludes by returning to the question of criteria (for selecting content) through raising the question 'what is basic?'.

Alternative emphases for science and technology education

Table 8 illustrates how science (A) and technology (B) can be combined separately with three typical curriculum emphases: the first emphasis is focused on teaching students about the nature and structure of the discipline (1); the second is aimed at the students' acquisition of skills corresponding to the processes used in each discipline (2); the third concerns the relationship of the subject to its broader social and economic context (3).

TABLE 9. Curriculum emphases in science and technology

Curriculum emphasis	Discipline	
	A: Science	B: Technology
1. Structure of the discipline	A1: Nature of science	B1: Nature of technology
2. Process skills	A2: Scientific skills	B2: Technological skills
3. Relationship with society	A3: Science and society	B3: Technology and society

At different times, all of these combinations have been in fashion with educators and curriculum developers. Currently, the most commonly found emphasis for technology education in schools around the world is the technological skills emphasis (B2). In science education, the emphases espoused by ministries of education are frequently a blend of A1, A2 and A3. However, classroom realities are often better represented by an emphasis (not shown here) that Roberts calls the 'correct explanations emphasis' [64, p. 247]. He characterizes it as the 'familiar "master now, question later" emphasis' [p. 248] that places stress on the students learning the subject matter as taught. It is the emphasis that is, perhaps, most at variance with the concept of 'education through science'.

Cells A3 and B3 represent a newly emerging emphasis on science, technology and society; this places most emphasis on the interdisciplinary relations of both science and technology. With these emphases, students are expected to understand how both science and technology influence and are influenced by their social and economic contexts. Such emphases contrast strongly with the introspective structure-of-the-discipline emphases (A1 and B1) that became popular in the curriculum reform movement of the 1960s. It was during this latter period that such an emphasis in science education became linked with the psychological fashion of 'learning by discovery' to produce 'science as discovery', in which it was intended that students would be initiated into the intellectual processes of the scientist. They would thus emerge as scientific thinkers having acquired both a knowledge of scientific theory and also skill at scientific processes.

Recent criticisms of science and technology curricula

Such an approach to science education has come in for serious criticism in recent years. For example, David Layton, writing about the British and American projects of the 1960s, many of which adopted such a discovery approach, comments as follows:

Only on the basis of the most superficial analysis of the nature of scientific activity, could it be said that such methods enable children to think and work in ways characteristic of a successful practitioner of science.... The view of science which informed the original Nuffield schools science project and the American ... study projects, was 'science as search' and in so far as these courses were concerned to put children in the position of a scientific inquirer after new knowledge, the stereotype was that of the pure scientist, supremely oblivious to the applications and wider implications of his work. Science was studied as an end in itself [41, p. 175, 177].

This criticism has been borne out by extensive analysis of some of the implicit messages ('curriculum emphasis') in textbooks commonly used in Canada in recent years. The conclusion of this research was that textbooks often *espouse* the goal of teaching students 'the scientific method' but then let them only 'rediscover' or 'verify' accepted scientific laws and relationships.

The problem is that, in the majority of cases, neither the scientific question, nor the design of the experiment, nor the application of the results are matters for class discussion. Instead, students are locked firmly, from the beginning to the end of a laboratory session, in a predetermined step-by-step procedure. By following directions in a laboratory manual, they learn to do what is expected of them, not to experience the real scientific method [72, p. 30].

This research has led one researcher to develop the concept of 'textbook science' to describe the pale reflection of 'real science' that is presented in schools [75].

Not only does the science that is taught in schools poorly reflect the scientific enterprise, it tends to ignore the cultural and economic context in which that enterprise operates. Within a country that is relatively advanced in science and technology, such a feature merely results in an oversimplified version of the roles of science and technology in society, itself a serious enough problem for such a country. In a developing country, however, where the importation of technology from industrialized countries is commonplace, this context-free approach can have serious consequences for students who are trying to understand science and technology in relation to the state of development of their own country. (For more on this topic, see King [38].)

A specific illustration may serve to clarify this point. To a certain degree, Canada is a net importer of curriculum ideas and a conclusion of the Science Council of Canada's recent study of science and technology education in that country was that science needs to be taught 'in a Canadian context' [72, p. 39-41]. By this, it meant that children in Canadian schools should learn how science and technology have contributed to the shaping of their country. They should learn about the men and women whose scientific and technological achievements have been landmarks in Canada's scientific heritage. Morover, the council concludes:

...although scientific knowledge is international in nature, it is supported economically, produced, applied and taught in a specific national context. Canada has many unique features, such as its northern latitude and cold climate, the long distances between centres of population, and its abundant natural resources. All of these features make it different from other industrialized countries. On the local level, these characteristics provide many opportunities to explore the science-technology-society interaction [72, p. 40].

The Council's report speaks, of course, to Canadians. But its point is also more general. The science and technology education of any country should be related to the cultural, social and economic context of that country.

What is basic?

One of the problems in selecting science content for teaching is the rate at which new scientific knowledge is being generated. Older knowledge can become obsolete and irrelevant. What then should be regarded as basic to science and technology education at the primary level?

The structure-of-the-discipline emphases tend to incorporate, as content, topics which are regarded by scientists as 'basic' to the discipline. For example, physical science courses tend to introduce at an early stage the idea of the particulate nature of matter, despite the abstract nature of this concept. The skills development emphasis, by contrast, places very little emphasis on specific content as long as activities that require specific skills can be conducted. Finally, the emphasis on science, technology and society incorporates a criterion stressing content that is important in the outside world.

Reflections and questions for deliberation 189

At the primary level of education, educators have traditionally been divided into those who have seen science as a process and those who have seen it as a body of knowledge [94]. An extreme version of the former has argued that we should not bother with specific content at all. What matters, according to proponents of this view, are the processes of science (the 'emphasis', to use our terminology). An example of an elementary school science programme based on this position is *Science—a process approach* developed by the American Association for the Advancement of Science in the late 1960s. While this course is very popular amongst some educators, its almost complete lack of specific content makes it somewhat inflexible when used by teachers who do not have the background to adapt it to local conditions.

The other approach is to base one's curriculum on the substance of science. For example, one popular approach is to focus on what some people have called the 'big ideas' of science, major abstractions such as matter and organism (for example, from the *Science curriculum improvement study*, another United States' project from the same period), abstractions that appear and reappear in many areas of scientific thought and have many more specific implications. While this approach commends itself to those who are worried about the obsolescence of specific scientific information, its problem lies in the difficulty which children have in accessing abstract concepts such as these.

The problem for curriculum developers, it seems, is in deciding what is truly *basic*. The traditional way of dealing with this is in terms of a pyramid of dependent knowledge with mathematics at the top, physics and chemistry just below, followed, at the next level, by the life and earth sciences, with the applied sciences (including agriculture, forestry, engineering and medicine) at the bottom. The authors of this model, Wynne-Edwards and Neale, recommend that the time has come for these ideas to be reversed:

We have defined the apex of the pyramid as the basis of science, and the base of the pyramid as its derived and dependent end.... We have chosen to drive the pyramid of science into the heads of school children point first, with a stiff preliminary dose of 'basic' mathematics and sciences taught in isolation. The result is so alienating to most children that only a few persist to become scientists or engineers (in which case they might just rediscover the 'real' world around them at last) [93].

The approach which reverses the traditional pyramid might be more successful at helping in the struggle to universalize science and technology education. This views the same model as a pyramid of experience starting from a practical level at which children encounter science and technology naturally, and moving, as they progress through their education, towards levels of greater abstraction. The 'process *versus* content debate' is avoided and *relevance* becomes the criterion for determining what is truly basic.

A small extension of this model sees technology education and science education as having important connections, reflecting the connections between science and technology in the real world. Yet in the classroom, the connections are few. James Rutherford has identified the progenitors of technology education as technical training and science education [83]. If one accepts this to be the case on theoretical grounds, then one is forced also to conclude, on the practical evidence, that technology education has been raised in a single-parent family. Until the advent of recent projects in technology education, such as the Schools Council Technology Project (in the United Kingdom), technology as a subject had been confined almost entirely to training in practical skills. And, as such, it lacked the prestige of the more academic subjects such as science, at least in the schools of the British and American traditions (see [43] for an account of the struggle for respectability of technology education in the United Kingdom during the past forty years).

As we noted earlier, technology as well as science can be taught with a variety of emphases. Potential content for technology education can be developed not only from the traditional craft and manual skills, but also from an understanding of the nature and origins of different technologies in every national context and from the implications of decisions concerning the progress of technological development. If the purpose of technology education is education through technology, then every country can teach much about its own history and culture through study of its indigenous technologies.

Questions for deliberation

17. Which emphases are regularly incorporated into science education and technology education?
18. How accurate is the image of science and technology presented in the curriculum?
19. How accurate an image is presented of the roles of science and of technology in society?
20. What criteria are used for the selection of content for science and technology education?
21. To what extent are local examples of science and of technology incorporated in the curriculum and for what purpose?
22. Are science and technology presented in a national context?

THE DELIVERY SYSTEM IN PRIMARY SCHOOL

Teaching methods

The essence of all effective teaching, in science as in other subjects, is the relationship between teacher and student. Teachers who are bright, competent, excited by the subject matter and sensitive to the needs of the students are the ones who can work miracles, even in the absence of ideal conditions for teaching. Correspondingly, teachers who are uncomfortable with their own background knowledge or who are demoralized because of the working conditions can negatively influence students' attitudes towards science.

It is therefore appropriate to consider teachers and their teaching methods as the central elements in the 'delivery' of science and technology education in schools. And, given the variety of goals for primary education that were outlined in Chapter III, one might expect a corresponding array of methods for teaching to be described here. Yet the somewhat depressing conclusion of the section on teaching methods in Chapter V is that 'the traditional teacher-centred delivery is still common and that knowledge is a main aim...[and] methodological renewal is rare or slow.'

Why should this be so? Are teachers unable or unwilling to adopt different approaches than the traditional 'chalk and talk'? Is the curriculum emphasis concept merely a theoretical fantasy which cannot be made operational in real schools? Or is change possible but elusive? Do new approaches to implementing innovations in teaching have better prospects of success? These are questions that clearly concern most Member States as they change and develop new goals and expectations for schools. This section and the next one address these problems briefly in the light of two recent research studies, one a set of case studies of science teachers at work and the other a theoretical argument for a new approach to teacher education.

In 1981 and 1982, a series of eight case studies in science education were conducted in schools in eight different locations in Canada [51] as part of a larger study, by the Science Council of Canada, into how science education in Canada could be improved [72]. These case studies were designed to help the council understand why, despite many attempts at curriculum reform, classroom events and teachers' methods remained so firmly entrenched in traditional approaches. The eight studies had obvious differences which corresponded not only to the different locations and types of school but also to the different researchers who conducted them. But, beyond the superficial differences, there was a core of similarity which derived from the fact that schools everywhere have certain features in common.

The study co-ordinators, Olson and Russell, have summarized some of the features of this core in an overview section which they have entitled 'Dynamics of change and dilemmas of practice'. Their comments on the schools in the Canadian study seem sufficiently appropriate to the data in Chapter V that we have quoted them at some length:

> Not all... teachers are trained scientists and not all work with ample resources, but all of them do work with large numbers of children whose abilities vary considerably and whose home support varies even more. Teaching children with such a range of social and psychological backgrounds is very demanding. Add to this difficulty the lack of any clear consensus about what schools are for and the result is a task that is ambiguous and poorly delineated. We believe that teachers actively counter these forces, which place unlimited demands on them, by interpreting and carrying out their jobs in a particular way. Given the uncertainties that exist about subject-matter competence, students' behaviour and educational goals, it is not surprising to us that teachers approach their work in ways that make it less uncertain.... To ask teachers to change their methods and objectives without first considering the reasons they behave as they do in the first place is unwise, to put it mildly [51, p. 28-29].

They go on to identify a series of practical dilemmas that teachers face and resolve through their daily routine. The following is a sample of these dilemmas:

— How can teachers include science in the early years when society demands the teaching of 'basics'?
— How can teachers control students' energies without suppressing their imagination¿
— How can teachers reconcile the apparent objectivity of science with the apparent subjectivity of value-laden issues related to science?
— How can teachers meet the expectations of parents and students for grades and credentials while at the same time pursuing sidelines that are not directly related to testing and examination? [51, p. 29].

Dilemmas such as these are the answer, in our view, to the questions about the apparent lack of change. And, while they were formulated on the basis of evidence from Canadian schools, they may reflect aspects of others' experiences also.

How can such dilemmas be resolved in order that change can happen? To deal with this question one must recognize that, while the pedagogical dilemmas that face teachers the world over may be similar, each teacher in each classroom must work out his or her resolution to them. Generalized responses cannot be made. One must therefore look not for a set of panaceas but for a process through which each teacher separately can resolve these dilemmas in the context of the unique teaching situation. And for clues to this process, we wish to draw readers' attention to the work of Schön in his book *The reflective practitioner* [70].

Teacher education

Teacher training has gone through two major paradigms since it became formalized about one hundred years ago. The first of these, which is still found in many institutions, was based on the transmission of the skills and techniques of an experienced or 'master'-teacher to the novice largely by means of direct imitation. The 'model lesson' concept was characteristic of such a training approach. The master-teacher would plan and teach a lesson and the pupil-teacher would then teach the same lesson while being observed by the master, who would then offer correction and advice. There was little of what would now be called 'theory', though some psychology would be taught, as well as rules for school administration, health and safety, and so on. In addition, since most teachers at the primary level had not gone beyond the level of high school education themselves, lectures in subject matter content, such as science, would also be provided.

A radically different approach began to become common during the 1950s and 1960s in many parts of the world and particularly in those countries in which new curricula for science and technology education were being developed. These new curricula went far beyond the old textbooks, which had merely contained scientific information. They embodied a new approach to teaching based on a particular philosophy of science, a particular philosophy of education, and a particular psychology of learning. Students in teacher training were therefore subjected to quantities of the 'theory' of these new courses. As well, they were drilled in the new teaching techniques that represented the 'application' of this theoretical approach. Finally, they were introduced to the courses themselves; these embodied all the pedagogical virtues to which the students had been introduced.

The difference between the first and second paradigm is clear. As research in the pedagogy of science education increased, so this 'theory' became applied through new curricula. The term 'teacher-proof curricula' was even used to stress the view that the teacher was merely a channel for a complete science education package, itself scientifically designed, to be 'delivered'. Teacher training was the means by which it was ensured that teachers did not interfere with this delivery system. By contrast, the earlier paradigm saw the teacher as central to the educational process and teacher training was the means by which teachers would acquire the knowledge which had to be transmitted and the teaching techniques which had been proven by experience.

This is not the place in which to offer a thorough critique of these two models of the teacher training process. The weaknesses of the first have been discussed in the literature surrounding the introduction of new curricula in the 1960s, while the weaknesses of the second are becoming apparent only recently. Schön has conducted studies of various professionals at work, and has demonstrated that 'professional knowledge' is not simply *applied* scientific knowledge but what he calls 'reflection in action'.

With this phrase, Schön captures the essentially improvisational way in which, for example, a teacher collects, assimilates and synthesizes information about his or her teaching situation and plans appropriate actions based on his or her previous experience. These thought processes are often both intuitive and instantaneous, with the result that individuals find it difficult to articulate why, precisely, they chose to act in a particular way in a particular situation, even though it might be generally agreed that the action was wisely chosen. The better teacher is one who has learned to be reflective in the context of action.

Adopting this way of thinking about the role of teachers would radically alter approaches to teacher training. Instead of asking teaching trainees to adopt (mindlessly) the techniques of another teacher or (equally mindlessly) the 'correct' methods embodied in a curriculum package or even to learn enough theory and apply it rigorously, one is inviting teachers to develop personal judgement based on disciplined and self-conscious reflection. Such an approach to teacher training has yet to be advocated widely, but it would appear to represent a new and significant paradigm for this important enterprise.

The present book is concerned especially with science and technology education and, in conclusion, we offer some comments on teacher education in relation to these disciplines. At stake, as we have noted earlier, is the achievement of the multiple goals of education through science and technology. And the mechanism by which this can be implemented is the curriculum emphasis concept. What, we may therefore ask, is the relationship between the various emphases and the three paradigms of teacher education described here?

The first paradigm is essentially conservative, in that it encourages teachers to follow closely the methods and techniques of those who have gone before. While these may be excellent, such an approach toward teacher education is unlikely to stimulate the introduction of new curriculum emphases. A master teacher who teaches science with a 'correct explanations' emphasis is likely to insist that his student adopt the same emphasis. Thus, the introduction of new emphases to accommodate the changing times and changing goals is likely to be thwarted.

The second paradigm is no more flexible. The emphasis that was built into the curriculum projects of the 1960s was 'sold' to educators not as having been *selected* to suit national needs but as being *right* in a more absolute sense because it was supposedly 'based on research'. Teacher education tended therefore to be a process of indoctrinating student teachers into accepting as desirable the particular emphasis then in fashion.

The third paradigm is fundamentally different in that it credits the teacher with the ability to make informed judgements about the teaching situation and tries to stimulate that ability. In particular, it could be used to assist teachers to consider the consequences for students of their selecting one or another teaching technique and thus coming to see alternative approaches to teaching science as means to different ends rather than as better (or worse) means to the same end. The notion that students need to be exposed to a variety of different curriculum emphases in order to achieve the full range of objectives of science education is consistent with this paradigm for teacher education but scarcely with the others.

Instructional resources

About material resources for teaching we shall have little to comment, except to draw again on the concept of curriculum emphases and to reflect on the different approaches to teaching that each different emphasis implies.

If the goal of instruction is to have students acquire the skills of the scientist, as was the case with many of the primary school programmes of the 1960s, it is essential that laboratory-style experimental work be carried out. Now, as we have noted, critics have shown that the 'experiments' typically given to students to carry out these programmes are often a pale reflection of real scientific experiments. Nevertheless, they required that scientific equipment be available for students to use. When these new programmes were first introduced in their countries of origin, this meant a major capital expenditure by the schools on science laboratories and equipment. Since then, it has caused many other countries either not to implement science programmes, or to use substitute home-made equipment or to have the teacher demonstrate an experiment rather than enabling the students to carry it out themselves. Clearly, this distorts the intent of the original curriculum, but it is a necessary constraint where resources are limited.

The problem is that, over the past twenty years since programmes with this type of emphasis have been in fashion, science has become associated with great expense and therefore with schools that can afford scientific equipment. However, if one were to change the emphasis from scientific skill development to one of science in society, then the sort of student activity will also change. It may not be necessary to have students conduct laboratory-style experiments. Instead, it might be more appropriate to have them survey the ways in which electricity is used within the school building or at home in order to determine wastage. Such an activity could be in the context of a unit on forms of energy with a science-in-society emphasis (the example is taken from an actual unit developed for 13-year-olds [53]). The equipment needed for teaching such a unit is not only less in quantity and cost, it can also be found around the school, home or community with relative ease. Thus a decision to change emphasis can have major implications for the type and cost of the equipment needed. Only when the teaching of science becomes locked into one particular emphasis does the purchase of particular equipment become a prerequisite for science teaching.

The same consideration applies to the use of audio-visual aids and communication tools such as films, television and so on. Teachers need to ask themselves the purpose of using certain types of instructional aids such as these. While films are often motivating for students and a relaxation from real teaching for the teacher, they can also be an integral part of the teaching approach. For example, the same unit on 'Energy' from which the previous example came also requires that students visit a power station and, preferably, a hydro-electric generator. Of course, many schools will not have such a facility nearby and in such a situation a film or video-cassette can bring the experience to the students. The educational objectives of the unit make this an integral component of the emphasis.

Evaluation

The final aspect of the delivery system, which is of crucial importance, is the system of evaluation employed. It is our experience that inappropriate methods for evaluating student achievement do more to distort the achievement of the objectives of education *through* science and technology than any other factor. Too often, an elaborate effort is made to incorporate a particular emphasis into the science programme with the intention that students will learn both the subject matter and also achieve other objectives, yet only the learning of subject matter is tested. For example, a teacher could set out with the intention of teaching about the properties of matter (the science content) *and* about the nature of inference in science. However, if the test at the end of the unit only asks students to recall their knowledge of the properties of matter and ignores their understanding of inference, they will 'learn' the general lesson that 'only the science content matters'. The 'emphasis stuff' which, in this case, is intended to contribute to the broader educational objective of developing thinking skills is disregarded and appears to be of no value.

If, therefore, all the goals of science and technology education are to be taken seriously by teachers and by students, then those responsible for setting examinations must, in turn, devise ways in which to assess students' achievements of all the goals. This will take a major effort in many countries where traditional testing has been of a narrow content variety in the past. There are some good models of tests that attempt to assess non-content goals, but these must be sought out and built upon if a broader range of goals for science and technology education is to be achieved.

Reflections and questions for deliberation   195

Questions for deliberation

These follow the four sub-sections of this part of the chapter: teaching methods, teacher education, instructional resources and evaluation.

(a) Teaching methods:

23. What are the practical dilemmas faced by teachers in the classroom and how can they be resolved?

24. What alternative teaching methods are practicable within the context of our schools?

(b) Teacher education:

25. Which paradigm governs the system of teacher education in our universities or colleges? Are alternatives considered?

26. Are methods for teaching a variety of emphases taught as part of our teachers' training, or only one or two?

27. How can teachers be assisted to reflect in action?

(c) Instructional resources:

28. Is the equipment used for teaching science suited only to one curriculum emphasis or to a variety?

29. What new equipment or resources would be required if different emphases were introduced into the curriculum?

30. To what extent are films, field trips and other activities used to reach particular educational goals or are they used mainly to motivate students?

(d) Evaluation:

31. Are the tests that are set for students in science and technology education appropriate for the full range of objectives? If not, which ones are stressed or ignored?

32. How can the objectives of science and technology education other than knowledge recall be assessed?

RENEWAL OF SCIENCE AND TECHNOLOGY EDUCATION

In this final section, we shall not comment specifically on the data from Member States presented in Chapter VI concerning the goals and processes of the renewal of science and technology education. Rather, we shall draw our conclusions based on our overall study and expressed in the terms with which the study began (in Chapter I). This section is in two parts, the first of which reviews the state of the partnership among science, technology and education. This part implicitly sets out goals for renewal. The second part draws from our own experience concerning the process of renewal in order to suggest ways to 'renew' the process itself.

The state of the partnership

This book began with an argument that science, technology and education should function as *partners* in the struggle for social and economic development. It reviewed the components of that partnership — technology and education, technology and science, science and education — and concluded by asking about the extent to which the partnership was functioning effectively in the world today.

The evidence submitted by Member States to the International Conference on Education is both encouraging and discouraging. On the other hand, the ideals for education through both science and technology, as these are articulated by Member States, show that science in school is no longer perceived as an elitist subject intended for the few who will go on to study science at university. The universalization both of primary education in general and of science education in particular reflect the (now) commonly held view that all can and should benefit from a scientific education.

On the other hand, the substance of science education has, in many places, yet to be adapted to this new view. The result is that students do not see the relevance of science to their society, to their environment or to the world of work. In the analysis presented in this chapter, we have articulated this problem in terms of two principal concepts: levels of the curriculum; and curriculum emphases. Using these concepts, we can summarize our conclusions as follows.

At the level of the intended curriculum — that enshrined in policies by Ministries of Education — a broad range of goals (and thus of curriculum emphases) are called for in primary education. These are somewhat more limited when the goals of science and technology education are specified, but still more is expected of this sector of education than the mere learning of scientific information. Three emphases were used to represent the range of goals frequently specified for science and technology education, one in which students were supposed to understand the nature of science and technology, one in which they were expected to acquire the skills of the scientist and technologist, and one in which they would gain an understanding of the relationships among science, technology and society.

At the level of the planned curriculum and the taught curriculum, the evidence we reviewed indicated that this range of emphases was severely narrowed. Constraints of a practical nature that affect everyday life in schools in most parts of the world make it impossible to achieve all of the goals that are intended for science and technology education. But, in addition, such factors as teacher education and systems of evaluation, which are not simply the consequence of inadequate resources, can also function as constraints on the extent of the science and technology education that students receive. Evidence concerning the learned curriculum was beyond the scope of this study. However, the International Association for the Evaluation of Educational Achievement (IEA) is currently conducting a study of science learning in over forty countries of the world which, when completed, will provide some evidence of what is actually learned in science at school.

Ironically, the emphasis that is likely to do the most to support the partnership between science, technology and education is the one that is also the most likely to be absent from school science and technology programmes. The science/technology/society emphasis is intended to help students appreciate how science and technology affect society both in day-to-day practical ways — wastage of electricity was used as an example — but also in political, social and economic ways — in medicine, agriculture or communications, for instance. There was little evidence that such an emphasis forms a major part of science and technology programmes anywhere in the world, although experimental and new courses based on some of these ideas are beginning to appear in some countries.

In most places, however, it would appear that traditional approaches to science education — in which children are expected exclusively to learn the facts and principles of science — are the norm. Similarly, traditional approaches to technology education, in which children are trained to perform the manual tasks associated with traditional industrial technologies, are commonplace. It should be stressed that, in reporting this, we are in no way suggesting that students should not learn scientific information nor become technically skilled. Rather we are commenting on the difference between the achievement of these goals and the achievement of the range of goals called for by Ministries of Education *through* science and technology education.

In the course of this chapter, we have also raised questions about what can productively be achieved in and by schools and what aspects of education are better carried out in the home or workplace. The renewal of science and technology education requires us to recognize that education can and does take place elsewhere than in school and that more than just the institution of school be part of the renewal process. And it is to this process that we now turn for a final comment.

Renewing the renewal process

The need to renew science and technology education stems from the evident gaps between the various levels of the curriculum — intended, planned, taught and learned. Correspondingly, the goal of the renewal process must be to close these gaps as much as possible.

The method for introducing educational renewal most often recommended in textbooks on educational (and particularly curriculum) change involves a management model in which the goals of renewal are *first* determined, *then* the means for reaching these goals are established, and *then* those responsible for implementing these decisions at the next level of the system are brought into the process.

There is an undeniable logic to such a model but the literature is also full of instances in which it simply has not worked in practice. The problem is that educational innovation is, as House has pointed out [23], not a technology but a craft. It is a process in which people must make professional judgements about unique situations at every level. And this fact can be the downfall of the renewal process — but also the key to its success.

Research into this problem by one of the authors [52] has convinced us that, to be effective, the curriculum policy-making process must achieve two goals simultaneously: it must yield a rational product (a coherent curriculum policy); and it must stimulate action on the part of those responsible. There is a difference between *determining the answer* to a policy problem and *resolving* it, and that difference affects whether or not the intended renewal actually results in practice.

This conceptual research was tested in practice by a large-scale renewal project in Canadian science education which has been referred to earlier [55, 56]. The methodology, called 'deliberative inquiry', is based on the idea that successful renewal requires three essential ingredients: *people* committed to deliberation and the possibility of change; *information* about the context in which changes will take place; *ideas* concerning alternative directions and methods for change. Deliberative inquiry is a strategy for integrating these three elements into a process of renewal.

The key component of the strategy is that of 'deliberation' about both the goals and methods of the renewal by *all those affected* [71]. Thus, for example, in the Canadian study, scientists, engineers, employers, labour unions, parents and students participated in conferences on the science curriculum, as well as Ministries of Education officials, school administrators, teachers and university professors. The conclusions of these deliberations were endorsed by all of these groups and this has resulted in greater understanding and co-operation in the period that followed.

In deliberative inquiry, all three essential elements — people, information and ideas — must be blended together harmoniously. If the prejudices of individuals are allowed to deny the facts or to stifle the ideas, then the outcome will not be acceptable to all.

And, in such a circumstance, those who feel that their interests were not taken into account will find a way to block the renewal process.

It is important that this *deliberative* process should not be regarded as a denial of the *decision-making* prerogatives of those in positions of authority or responsibility. After deliberations have reached consensus on the goals and means of renewal, it is still the job of the formally designated officials to decide what shall be done. In this process they have other factors (such as financial priorities) which must be taken into account, as well as the *advice* they have received from the deliberative process. However, a decision consistent with the conclusions of deliberations by many individuals is more likely to be successfully implemented than one which is not.

At present, there are both national and international levels at which such deliberations already take place. Yet, how representative are these deliberations of all of the *shareholders* in science and technology education? To what extent are scientists, employers, labour unions, economists and policy makers in adjacent fields involved in policy making for science and technology education as well as science educators? If science, technology and education are to become effective partners in the struggle for national development, then it is perhaps time that we who are educators begin to see ourselves as part of a larger system. Science, technology and education are at the heart of economic and social change in the world today. Policy making in these three fields must therefore reflect this reality and become an integrated process.

References

BIBLIOGRAPHIC REFERENCES
1. **Aikenhead, G.** Science at Prairie High School. In: **Olson, J.; Russel, T.**, eds. *Science education in Canadian Schools, volume III: Case studies of science teaching.* Ottawa, Supply XY Services Canada, 1984, p. 257-289. (Science Council of Canada. Background study, 52)
2. **Anderson, R.C., et al.** *Current research on instruction.* Englewood Cliffs, NJ, Prentice-Hall, 1969. 396 p.
3. **Black, P.J.** *Science in primary education.* Strasbourg, Council for Cultural Co-operation, Council of Europe, 1984. 21 p. (DEICS/Rech (84) 37) ((Paper prepared for the Educational Research Workshop on Science in Primary Education, Edinburgh, 1984]
4. **Chaïbderraine, M.** *Educational wastage at the primary level.* Linkoping, Sweden, Department of Education, Linkoping University, 1978. 205 p. (Linkoping studies in education. Dissertations, no. 9)
5. **Collins, S.** The social and economic causes of wastage in schools and other educational institutions in Tanganyika. *Teacher education* (London), vol. 5, no. 1, May 1964, p. 40-50.
6. **Conference of Ministers of Education and those Responsible for Economic Planning in African Member States, 5th, Harare, 1982.** *Final report.* Paris, Unesco, 1982. 94 p. (ED/MD/69) [Unesco microfiche 83s0003]
7. **Conference of Ministers of Education of Member Stats of the Europe Region, 3rd,** Sofia, **1980.** *Final report.* Paris, Unesco, 1980. 82 p. (ED/MD/61) [Unesco microfiche 81s0171]
8. **Consultation Meeting on Pilot Project 'Technology in General Education', Beijing, 1984.** *Final report.* Paris, Unesco, 1984. 18 p. (ED.84/WS/86) [Unesco microfiche 84s0741]
9. **Cowen, R.** *International yearbook of education. Volume XXXIV - 1982: Educational structures.* Paris, Unesco, 1983. 159 p.
10. **Dahllöf, U.** *Skoldifferentiering och undervisningsförlopp: komparativa mål- och processanalyser av skolsystem, I.* Stockholm, Almqvist XY Wiksell, 1967.
11. **D'Silva, E., ed.** *Education: sector policy paper.* 3rd ed. Washington, DC, World Bank, 1980. 143 p.
12. **Elgvist-Saltzman, Inga.** *Vagen genom univresitetet: en forskningsöversikt och en empirisk analys av några studieväger inom filosofisk fakultet i Umeå, 1968-1975.* Uppsala, Sweden, Department of Education, University of Uppsala, 1976. 223 p. (Uppsala studies in education, 1)
13. **Emanuelsson, I.** *Studieavbrott i grundskolan 5.* Stockholm, Pedagogiskt centrum vid Stockholms skoldirection, 1976. 20 p. (Pedagogiskt utvecklingsarbete vid Stockholms skolor, nr 66)
14. **Eriksson, K.H.** *Fågel eller fisk: Barn—ungdom—naturvetenskap—teknik. Kunskapsöversikt och probleminventering: rapport över*

BUNT-projektet. Linköping, Sweden, Department of Education, University of Linköping, 1984.

15. **Gage, N.L.., ed.** *Handbook of research on teaching.* Chicago, IL, Rand McNally, 1963. 1218 p.

16. **Girod, R.** *Inégalité, inégalités.* Paris, Presses universitaires de France, 1977.

17. **Great Britain. Central Advisory Council for Education (England).** *Children and their primary schools.* London, Department of Education and Science, 1967. 2 v. [The Plowden report]

18. **Guthrie, J.W.** A survey of school effectiveness studies. *In:* **United States Office of Education. Bureau of Educational Personnel Development.** *Do teachers make a difference?* Washington, DC, 1970, p. 25-54.

19. **Harlen, Wynne, ed.** *New trends in primary school science education, volume I.* Paris, Unesco, 1983. 216 p.

20. **Harlen, Wynne.** Does content matter in primary science? *School science review* (Hatfield, United Kingdom, Association for Science Education), vol. 59, no. 209, June 1978, p. 614-625.

21. **Heyneman, S.P.** Relations between teachers' characteristics and differences in academic achievement among Ugandian primary schools. *Education in Eastern Africa* (Nairobi, Regional Council for Education), vol. 6, 1976, p. 41-51.

22. **Holmes, B.** *International yearbook of education. Volume XXXV-1983: Educational development trends.* Paris, Unesco, 1983. 232 p.

23. **House, E.R.** Technology versus craft: a ten year perspective on innovation. *Journal of curriculum studies* (London), vol. 11, no. 1, January-March 1979, p. 1-15.

24. **Al-Huqail, S.A.** *Educational policy in the Kingdom of Saudi Arabia: its principles, objectives, means of implementation, directions and examples of achievement.* Riyadh, Research Center, Imam Muhammad Ibn Saud Islamic University, 1404 AH (AD 1984).

25. **Husén, T.** *The school in question: a comparative study of the school and its future in Western society.* London, Oxford University Press, 1979. 196 p.

26. **Husén, T.** *Utbildning i internationellt perspektiv.* Stockholm, Liber Läromedelsförlaget, 1985.

27. **Husén, T., ed.** *International study of achievement in mathematics: a comparison of twelve countries.* Vols. I - II. Stockholm, Almqvist & Wiksell; New York, Wiley, 1967. 304, 368 p.

28. **Husén, T.; Saha, L.J.; Noonan, R.** *Teacher training and student achievement in less developed countries.* Washington, DC, The World Bank, 1978. 135 p. (World Bank. Staff working paper, no. 310)

29. **International Conference on Education, 38th Session, Geneva, 1981.** *Interaction between education and productive work.* Geneva, Unesco, 1981. 33 p. (ED/BIE/CONFINTED 38/5; ED.81/CONF.205/COL.6) [IBE microfiche SIRE/01634]

30. **International Conference on Education, 39th Session, Geneva, 1984.** *Contribution of the international teachers' organizations to the debate on the special theme: Universalization and renewal of primary education in the perspective of an appropriate introduction to science and technology.* Paris, Unesco, 1984. 22 p. (ED/BIE/CONFINTED 39/Ref.4; ED.84/CONF.206/COL.1) [IBE microfiche SIRE/01924]

31. **International Conference on Education, 39th Session, Geneva, 1984.** *Education for all in the new scientific and technical environment and taking into account disadvantaged groups.* Paris, Unesco, 1984. 23 p. (ED/BIE/CONFINTED 39/3; ED.84/CONF.206/COL.10) [IBE microfiche SIRE/01918]

32. **International Conference on Education, 39th Session, Geneva, 1984.** *L'enseignement primaire à la lumiére des recommendations des conférences régionales des ministres de l'éducation.* Paris, Unesco, 1984. 23 p. (ED/BIE/CONFINTED 39/Ref.3;

ED.84/CONF.206/COL.6) [IBE microfiche SIRE/01923]

33. **International Conference on Education, 39th Session, Geneva, 1984.** *Evolution of wastage in primary education in the world between 1970 and 1980.* Paris, Division of Statistics on Education, Office of Statistics, Unesco, 1984. 67p. (ED/BIE/CONFINTED 39/Ref.2; ED.84/CONF.206/COL.3) [IBE microfiche SIRE/01922]

34. **International Conference on Education, 39th Session, Geneva, 1984.** *Final report.* Paris, Unesco: International Bureau of Education, 1984. 81p. (ED/MD/79) [IBE microfiche SIRE/01917]

35. **International Conference on Education, 39th Session, Geneva, 1984.** *Major trends in education: 1981-1983.* Paris, Unesco, 1984. 32p. (ED/BIE/CONFINTED 39/4; ED.84/CONF.206/COL.8) [IBE microfiche SIRE/01919]

36. **International Conference on Education, 39th Session, Geneva, 1984.** *A summary statistical review of education in the world, 1960-1982.* Geneva, Unesco, 1984. 80p. (ED/BIE/CONFINTED 39/Ref. 1; ED.84/CONF.206/COL.4) [IBE microfiche SIRE/01921]

37. **International Conference on Education, 39th Session, Geneva, 1984.** *Universalization and renewal of primary education in the perspective of an appropriate introduction to science and technology.* Paris, Unesco, 1984. 38p. (ED/BIE/CONFINTED 39/5; ED.80/CONF.206/COL.9) [IBE microfiche SIRE/01920]

38. **King, K.** Education, science policy, research and action: a review paper. *In:* King, K., ed. *Science, education and society: perspectives from India and South East Asia.* Ottawa, International Development Research Centre, 1985, p.1-52.

39. **King, K.** The pursuit of science and technology manpower in the 1980s: some conceptual problems. *In:* King, K., ed. *Science, education and society: perspectives from India and South East Asia.* Ottawa, International Development Research Centre, 1985, p.105-143.

40. **Kulkarni, V.G.** Universalization of education problems and remedial measures. *In:* King, K., ed. *Science, education and society: perspectives from India and South East Asia.* Ottawa, International Development Research Centre, 1985, p.144-160.

41. **Layton, D.** *Science for the people: the origins of the school science curriculum in England.* London, Allen & Unwin, 1973. 226p.

42. **Lindgren Hooker, B.** *Educational flow models, with applications to Arab statistical data.* Linköping, Sweden, Department of Education, Linköping University, 1975. (Linköping studies in education. Dissertations, no. 3)

43. **McCulloch, G.; Jenkins, E.; Layton, D.** *Technological revolution? The politics of school science and technology in England and Wales since 1945.* London, Falmer Press, 1985. 240p.

44. **Marklund, S.** *Skolklassens storlek och struktur: studier rörande elevernas kunskaper i relation till klassens storlek och homogenitet.* Stockholm, Almqvist & Wiksell, 1962.

45. **Marklund, S.** *The IEA Project: an unfinished audit.* Stockholm, Institute of International Education, University of Stockholm, 1983. 33p. (Report, 64)

46. **Marklund, S.** *Skolsverige 1950—1975.* Del 4: Differentieringsfragan. Stockholm, Liber Utbildningsförlaget, 1985.

47. **Meeting of Experts on the Incorporation of Science and Technology in the Primary School Curriculum, Paris, 1980.** *Final report.* Paris, Unesco, 1980. 24p. (ED.80/CONF.618/COL.4) [Unesco microfiche 81s0164]

48. **Munby, H.** Introduction. *In:* Munby, H.; Orpwood, G.; Russell, T., eds. *Seeing curriculum in a new light: essays from science education.* Lanham, MD, University Press of America, 1984, p.1-7.

49. **Munby, H.; Russell, T.** A common curriculum for the natural sciences. *In:* Fenstermacher, G.D.; Goodlad, J.I., eds.

Individual differences and the common curriculum. Chicago, IL, University of Chicago Press, 1983, p. 160-185. (National Society for the Study of Education. Yearbook, 82, part I)

50. **National Commission on Excellence in Education, USA.** *A nation at risk: the imperative for educational reform.* Washington, DC, U.S. Government Printing Office, 1983. 65 p.

51. **Olson, J.; Russell, T., eds.** *Science education in Canadian schools, Volume III: case studies of science teaching.* Ottawa, Supply & Services Canada, 1984. 297 p. (Science Council of Canada. Background study, 52)

52. **Orpwood, G.W.F.** The reflective deliberator: a case study of curriculum policymaking. *Journal of curriculum studies* (London), vol. 17, no. 3, July—September 1985, p. 293-304.

53. **Orpwood, G.W.F.; Roberts, D.A.** *Force and energy: a science and society approach.* Toronto, Canada, OISE Press, 1981. 49 p.

54. **Orpwood, G.W.F.; Souque, J.-P.** *Science education in Canadian schools, volume I: introduction and curriculum analyses.* Ottawa, Supply & Services Canada, 1984. 227 p. (Science Council of Canada. Background study, 52)

55. **Orpwood, G.W.F.; Souque, J.-P.** Toward the renewal of Canadian science education: II. Deliberate inquiry model. *Science education* (New York), vol. 69, no. 4, July 1985, p. 477-489.

56. **Orpwood, G.W.F.; Souque, J.-P.** Towards the renewal of Canadian science education: II. Findings and Recommendations. *Science education* (New York), vol. 69, no. 5, October 1985, p. 625-636.

57. **Peters, R.S.** *Education as initiation.* London, Evans Brothers, 1964. 48 p.

58. **Postlethwaite, T.N.** *School organization and student achievement: a study based on achievement in mathematics in twelve countries.* Stockholm, Almqvist & Wiksell, 1967. 146 p.

59. **Postlethwaite, T.N.; Lewy, A.** *Annotated bibliography of IEA publications (1962—1978).* Stockholm, University of Stockholm, 1979. 54 p.

60. **[Regional] Conference of Ministers of Education and Those Responsible for Economic Planning in the Arab States, 4th, Abu-Dhabi, 1977.** Final report. Paris, Unesco, 1977. 82 p. (ED/MD/50) [Unesco microfiche 79s0692]

61. **Regional Conference of Ministers of Education and Those Responsible for Economic Planning in Asia, 3rd, Singapore, 1971.** Final report. Paris, Unesco, 1971. 91 p. (ED/MD/20)

62. **Regional Conference of Ministers of Education and Those Responsible for Economic Planning in Asia and Oceania, 4th, Colombo, 1978.** Final report. Paris, Unesco, 1978. 110 p. (ED/MD/53) [Unesco microfiche 79s0691]

63. **Regional Conference of Ministers of Education and Those Reponsible for Economic Planning of Member States in Latin America and the Caribbean, Mexico City, 1979.** Final report. Paris, Unesco, 1980. 113 p. (ED/MD/58) [Unesco microfiche 80s0620]

64. **Roberts, D.A.** Developing the concept of 'Curriculum emphases' in science education. *Science education* (New York), vol. 66, no. 2, April 1982, p. 243-260.

65. **Roberts, D.A.** The place of qualitative research in science education. *Journal of research in science teaching* (New York), vol. 19, no. 4, April 1982, p. 277-292.

66. **Roberts, D.A.** *Scientific literacy: towards balance in setting goals for school science programs.* Ottawa, Science Council of Canada, 1983. 43 p. (Discussion paper, D83/2)

67. **Roberts, D.A.; Orpwood, G.W.F.** Classroom events and curricular intentions: a case study in science education. *Canadian journal of education/Revue canadienne de l'éducation* (Saskatoon, Sask., Canadian Society for the Study of Education), vol. 7, no. 2, 1982, p. 1-15.

68. **Rosenshine, B.** *Teaching behaviours and student achievement.* Slough, United Kingdom, National Foundation for Educational Research, 1971. 229 p.

(International Association for the Evaluation of Educational Achievement. Studies, no. 1)
69. **Schoenberger, Mary M.** Teaching science at Seaward Elementary School. *In:* **Olson, J.; Russell, T., eds.** *Science education in Canadian schools, volume III: case studies of science teaching.* Ottawa, Supply & Services Canada, 1984, p. 30-64. (Science Council of Canada. Background study, 52)
70. **Schön, D.A.** *The reflective practitioner: how professionals think in action.* New York, Basic Books, 1983. 374 p.
71. **Schwab, J.J.** The practical: a language for curriculum. *In:* **Westbury, I.; Wilkof, N.J., eds.** *Science, curriculum and liberal education.* Chicago, IL, University of Chicago Press, 1978, p. 287-321.
72. **Science Council of Canada.** *Science for every student: educating Canadians for tomorrow's world.* Ottawa, Supply & Services Canada, 1984. 85 p. (Report, 36)
73. Science education in Asia and the Pacific. *Bulletin of the Unesco Regional Office for Education in Asia and the Pacific* (Bangkok), no. 25, June 1984, 498, li p.
74. **Simmons, J.** Education, poverty and development. *In:* **Simmons, J., ed.** *Investment in education: national strategy options for developing countries.* Washington, DC, International Bank for Reconstruction on and Development, 1975, p. 147-213. (World Bank. Working paper, no. 196) [Also published separately as Bank staff working paper, no. 188]
75. **Souque, J.-P.** Science education and textbook science. *Canadian journal of education/Revue canadienne de l'éducation* (Saskatoon, Sask., Canadian Society for the Study of Education) [In publication, 1987]
76. **Stake, R.E.; Easley, J.** *Case studies in science education.* Washington, DC, United States Government Printing Office, 1977. 2 v. (National Science Foundation. Report SE 78-74)
77. **Studstill, J.D.** Why students fail in Masomo, Zaire. *Journal of research and development in education* (Athens, GA, College of Education, University of Georgia), vol. 9, no. 4, Summer 1976, p. 124–136.
78. **Stufflebeam, D.L., ed.** *Educational evaluation and decision making.* Itasca, IL, Peacock Pubs., 1971.
79. **Svensson, N.-E.** *Ability grouping and scholastic achievement: report on a five-year follow-up study in Stockholm.* Stockholm, Almqvist & Wiksell, 1962. 236 p.
80. **Thorndike, R.L., ed.** *Educational measurement.* 2nd ed. Washington, DC, American Council on Education, 1971. 768 p.
81. **Törnvall, A.** *Läraren och läroplanen.* Linköping: Linkoping Studies in Education, Dissertations, no. 16, 1982.
82. **Trempe, P.-L.** . Lavoisier: science teaching at an Ecole polyvalente. *In:* **Olson, J.; Russell, T.; eds.** *Science education in Canadian schools, volume III: Case studies of science teaching.* Ottawa, Supply XY Services, Canada, 1984, p. 209–256. (Science Council of Canada. Background study, 52)
83. **Unesco.** *Science and technology education and national development.* Paris, 1983/34. 197 p.
84. **Unesco.** *Technology education as part of general education: a study based on a survey conducted in 37 countries.* Paris, 1983. 34 p. (Science and technology education document series, 4) (ED.83/WS/52) [Unesco microfiche 84s0075]
85. **Unesco.** *Statistical yearbook/Annuaire statistique/Anuario estadistico, 1983.* Paris, 1983. 1 v. (various pagings)
86. **Unesco. Office of Statistics. Division of Statistics on Education.** *Survey of teachers' qualifications/Enquête sur la qualification des enseignants.* Paris, 1977. 82 p. (CSR-E-23) [Unesco microfiche 78s0043]
87. **Werdelin, I.** [A chapter on post-compulsory and adult education in Sweden] *In:* **Crosgrove, D.; McDonnel, C., eds.** *Postcompulsory education and training: an international survey.* [1985?]

88. **Werdelin, I.** *Concept formation and concept learning: report of the work in a six-year project.* Linköping, Sweden, Department of Education, Linköping University, 1974. 150p. (Linköping Studies in Education. Reports, no. 1)
89. **Werdelin, I.** *Handbook of educational research methods: research designs.* Linköping, Sweden, Department of Education, Linköping University, 1982. 498p. (Linköping studies in education. Reports, no. 7)
90. **Werdelin, I.** *Manual of educational planning, 4: quantitative aspects of educational planning.* Linköping, Sweden, Department of Education, (Linköping studies in education, Reports, no. 2
91. **Werdelin, I.** *Manual of educational planning, 9: evaluation.* Linköping, Sweden, Department of Education, School of Education, Linköping University, 1975. 188p. Linköping University, 1977. 332p. (Linköping studies in education. Reports, no. 4)
92. **Werdelin, I.** *Quantitative methods and techniques of planning.* Beirut, Regional Centre for Educational Planning and Administration in the Arab Countries, 1972. 344p.
93. **Wynne-Edwards, H.R.** Science and cultural future: the usefulness of the history of science and technology to decision makers. *In:* **Jarrell, R.A.; Ball, N.R.**, eds. *Science, technology and Canadian history.* Waterloo, Ont., Wilfrid Laurier University Press, 1980, p. 94–103.
94. **Young, B.L.** The selection of processes, contexts and concepts and their relation to methods of teaching. *In:* **Harlen, Wynne**, ed. *New trends in primary school science education, volume 1.* Paris, Unesco, 1983, p. 7-16.

References

QUESTIONNAIRE REPLIES

ALGERIA. MINISTÈRE DE L'ÉDUCATION ET DE L'ENSEIGNEMENT FONDAMENTAL
Généralisation et rénovation de l'enseignement primaire dans la perspective d'une initiation scientifique et technique appropriée. Alger, Ministère de l'éducation et de l'enseignement fondamental, 1983. 11, i l., table. Reply to questionnaire ED/BIE/CONFINTED/39/Q.1/83. (fre)
Microfiche: SIRE/01643

ANGOLA. MINISTÈRE DE L'ÉDUCATION
Développement de l'éducation en République populaire d'Angola, 1981-1983. Luanda, Ministère de l'éducation, 1984. ii, 37, 13 l., figs., tables. Bibl.: p. 37. Includes the reply to questionnaire ED/BIE/CONFINTED/39/Q.1/83. (fre)
Microfiche: SIRE/01644

ARGENTINA. NATIONAL COMMISSION FOR UNESCO
Generalización y renovación de la enseñanza primaria en la perspectiva de una iniciación científica y técnica apropriada. Buenos Aires, Comisión Nacional Argentina de Cooperación con la Unesco, 1983. 18 l., tables. Reply to questionnaire ED/BIE/CONFINTED/39/Q.1/83. (spa)
Microfiche: SIRE/01646

AUSTRALIA. NATIONAL COMMISSION FOR UNESCO
Universalization and renewal of primary education in the perspective of an appropriate introduction to science and technology. Canberra, Australian National Commission for Unesco, 1984. 31 l., tables. Bibl.: l. 30-31. Reply to questionnaire ED/BIE/CONFINTED/39/Q.1/83. (eng)
Microfiche: SIRE/01648

AUSTRIA. FEDERAL MINISTRY OF EDUCATION AND ARTS
Universalization and renewal of primary education in the perspective of an appropriate introduction to science and technology. Wien, Bundesministerium für Unterricht und Kunst, 1983. 6 l., table. Reply to questionnaire ED/BIE/CONFINTED/39/Q.1/83. (eng)
Microfiche: SIRE/01651

BAHAMAS. MINISTRY OF EDUCATION
(Universalization and renewal of primary education in the perspective of an appropriate introduction to science and technology). Nassau, Ministry of Education, 1983. 13 l. Reply to questionnaire ED/BIE/CONFINTED/39/Q.1/83. (eng)
Microfiche: SIRE/01653

BAHRAIN. MINISTRY OF EDUCATION. DIRECTORATE OF PLANS AND PROGRAMMING
(Taʿmīm at-taʿlīm al-ibtidāʾī wa-tagdīduhu bi-hadaf tadmīnihi qadran mulāʾiman min mabādiʾ al-ʿilm wa-ät-tiknūlūǧiyya). Manāma, Idārat al-ḥuṭaṭ wa-äl-barmaǧa, Wazārat at-tarbiya wa-ät-taʿlīm, 1983. 38 l., tables. Reply to questionnaire ED/BIE/CONFINTED/39/Q.1/83. (ara)
Microfiche: SIRE/01656

BANGLADESH BUREAU OF EDUCATIONAL INFORMATION AND STATISTICS
Universalization and renewal of primary education in the perspective of an appropriate introduction to science and technology. Dhaka, Bangladesh Bureau of Educational Information and Statistics, Ministry of Education, 1983. 15 l., tables. Reply to questionnaire ED/BIE/CONFINTED/39/Q.1/83. (eng)
Microfiche: SIRE/01659

BELGIUM. MINISTÈRE DE L'ÉDUCATION NATIONALE
Généralisation et rénovation de

l'enseignement primaire dans la perspective d'une initiation scientifique et technique appropriée. Bruxelles, Ministère de l'éducation nationale, 1983. 22 l., tables. Reply to questionnaire
ED/BIE/CONFINTED/39/Q.1/83. (fre)
Microfiche: SIRE/01662

BENIN. NATIONAL COMMISSION FOR UNESCO
(Généralisation et rénovation de l'enseignement primaire dans la perspective d'une initiation scientifique et technique appropriée). Porto-Novo, Commission nationale béninoise pour l'Unesco, 1984. 8 p., table. Reply to questionnaire
ED/BIE/CONFINTED/39/Q.1/83. (fre)
Microfiche: SIRE/01665

BOTSWANA. MINISTRY OF EDUCATION
Universalization and renewal of primary education in the perspective of an appropriate introduction to science and technology. Gaborone, Ministry of Education, 1984. 15 l. Reply to questionnaire
ED/BIE/CONFINTED/39/Q.1/83. (eng)
Microfiche: SIRE/01668

BRAZIL. MINISTÉRIO DA EDUCAÇÃO E CULTURA. SECRETARIA DE ENSINO DO 1O E 2O GRAUS
Generalização e renovação do ensino primário na perspectiva de uma iniciação científica e técnica apropriada. Rio de Janeiro, Secretaria de Ensino do 1o e 2o Graus, Ministério da Educação e Cultura, 1983. 11 l., table. Reply to questionnaire
ED/BIE/CONFINTED/39/Q.1/83. (por)
Microfiche: SIRE/01669

BULGARIA. NATIONAL COMMISSION FOR UNESCO
(Généralisation et rénovation de l'enseignement primaire dans la perspective d'une initiation scientifique et technique appropriée). Sofia, Commission nationale de la R.P. de Bulgarie pour l'Unesco, 1983. 30 l. Reply to questionnaire
ED/BIE/CONFINTED/39/Q.1/83. (fre)
Microfiche: SIRE/01672

BURUNDI. NATIONAL COMMISSION FOR UNESCO
Généralisation et rénovation de l'enseignement primaire dans la perspective d'une initiation scientifique et technique appropriée. Bujumbura, Commission nationale du Burundi pour l'Unesco, 1984. 8, 16 l., tables. Reply to questionnaire
ED/BIE/CONFINTED/39/Q.1/83. (fre)
Microfiche: SIRE/01675

BYELORUSSIAN SSR. MINISTERSTVO PROSVEŠČENIJA. UPRAVLENIE ŠKOL
(Vseobščee rasprostranenie i obnovlenie načal'nogo obrazovanija putem vvedenija sootvetstvujuščih elementov naučnyh i tehničeskih znanij). Minsk, Upravlenie škol, Ministerstvo prosveščenija, 1983. 27 l., tables. Reply to questionnaire
ED/BIE/CONFINTED/39/Q.1/83. (rus)
Microfiche: SIRE/01678

CAMEROON UR. MINISTÈRE DE L'ÉDUCATION NATIONALE. DIRECTION DE L'ENSEIGNEMENT PRIMAIRE ET MATERNEL. SERVICE DES AFFAIRES PÉDAGOGIQUES ET DE LA FORMATION
Généralisation et rénovation de l'enseignement primaire dans la perspective d'une initiation scientifique et technique appropriée. Yaoundé, Service des affaires pédagogiques et de la formation, Direction de l'enseignement primaire et maternel, Ministère de l'éducation nationale, 1984. 10 l. Reply to questionnaire
ED/BIE/CONFINTED/39/Q.1/83. (fre)
Microfiche: SIRE/01680

CANADA. COUNCIL OF MINISTERS OF EDUCATION (CANADA)
Education in Canada, 1981–

References

1983/*L'enseignement au Canada, 1981–1983.* Ottawa, Council of Ministers of Education, 1984. 1 v. (various pagings), figs., map, tables. Includes bibliographies. Part I: Education in the ten provinces of Canada, prepared by Council of Ministers of Education; Part II: The Government of Canada, prepared/coordinated by the Department of the Secretary of State; Part III: The Northwest Territories and the Yukon Territory, prepared by the Departments of Education of the Northwest Territories and the Yukon Territory. Includes the replies to questionnaires ED/BIE/CONFINTED/39/Q.1/83 and ED/BIE/CONFINTED/39/Q.2/83. Report submitted by the Secretary of State for External Affairs. (same text in eng, fre)
Microfiche: SIRE/01682

CENTRAL AFRICAN REPUBLIC. INSTITUT NATIONAL D'ÉDUCATION ET DE FORMATION (CENTRAL AFRICAN REPUBLIC)
(Généralisation et rénovation de l'enseignement primaire dans la perspective d'une initiation scientifique et technique appropriée). Bangui, Institut national d'éducation et de formation, 1983. 19 l. Reply to questionnaire ED/BIE/CONFINTED/39/Q.1/83. (fre)
Microfiche: SIRE/01683

CHILE. NATIONAL COMMISSION FOR UNESCO
Generalización y renovación de la enseñanza primaria en la perspectiva de una iniciación científica y técnica apropiada. Santiago, Comisión Nacional Chilena de Cooperación con la Unesco, 1983. 28 l. Reply to questionnaire ED/BIE/CONFINTED/39/Q.1/83. (spa)
Microfiche: SIRE/01685

CHINA. NATIONAL COMMISSION FOR UNESCO
(Universalization and renewal of primary education in the perspective of an appropriate introduction to science and technology). Beijing, National Commission of the People's Republic of China for Unesco, 1984. 10 l. Reply to questionnaire ED/BIE/CONFINTED/39/Q.1/83. (eng)
Microfiche: SIRE/01688

COLOMBIA. MINISTERIO DE EDUCACIÓN NACIONAL
(Generalización y renovación de la enseñanza primaria en la perspectiva de una iniciación científica y técnica apropiada). Bogotá, Ministerio de Educación Nacional, 1983. 39, 17 l., fig., tables. Reply to questionnaire ED/BIE/CONFINTED/39/Q.1/83. (spa)
Microfiche: SIRE/01691

CONGO. MINISTÈRE DE L'ÉDUCATION NATIONALE. DIRECTION GÉNÉRALE DE L'ÉDUCATION FONDAMENTALE. DIRECTION DE L'ÉDUCATION FONDAMENTALE PREMIER DEGRÉ. SERVICE DE L'INSPECTION ET DE L'ENCADREMENT PÉDAGOGIQUES
Réponses au questionnaire sur la généralisation et rénovation de l'enseignement primaire. Brazzaville, Service de l'inspection et de l'encadrement pédagogiques, Direction de l'éducation fondamentale premier degré, Direction générale de l'éducation fondamentale, Ministère de l'éducation nationale, 1984. i, 12 p., tables. Reply to questionnaire ED/BIE/CONFINTED/39/Q.1/83. (fre)
Microfiche: SIRE/01694

CUBA. MINISTERIO DE EDUCACIÓN
Generalización y renovación de la enseñanza primaria en la perspectiva de una iniciación científica y técnica apropiada. La Habana, Ministerio de Educación, 1983. 12 l. Reply to questionnaire ED/BIE/CONFINTED/39/Q.1/83. (spa)
Microfiche: SIRE/01696

CYPRUS. MINISTRY OF EDUCATION
The Universalization and renewal of primary education in the perspective of an appropriate introduction to science and technology. Nicosia, Ministry of Education, 1983. 12 l., tables. Reply to questionnaire
ED/BIE/CONFINTED/39/Q.1/83. (eng)
Microfiche: SIRE/01699

CZECHOSLOVAKIA. NATIONAL COMMISSION FOR UNESCO
Vseobščee rasprostranenie i obnovlenie načal'nogo obrazovanija putem vvedenija sootvetstvujuščih elementov naučnyh i tehničeskih znanij. Praga, Čehoslovackaja komissija po delam Yunesko, 1983. 34 l., table. Reply to questionnaire
ED/BIE/CONFINTED/39/Q.1/83. (rus)
Microfiche: SIRE/01702

DENMARK. MINISTRY OF EDUCATION
Universalization and renewal of primary education in the perspective of an appropriate introduction to science and technology. Copenhagen, Ministry of Education, 1984. 13 p., tables. Reply to questionnaire
ED/BIE/CONFINTED/39/Q.1/83. (eng)
Microfiche: SIRE/01705

EGYPT. NATIONAL CENTER FOR EDUCATIONAL RESEARCH (EGYPT). DOCUMENTATION AND EDUCATIONAL INFORMATION AGENCY
Universalization and renewal of primary education in the perspective of an appropriate introduction to science and technology. Cairo, Documentation and Educational Information Agency, National Center for Educational Research, 1983. ii, 54 p., tables. Reply to questionnaire
ED/BIE/CONFINTED/39/Q.1/83. (eng; also in ara)
Microfiche: SIRE/01709

ETHIOPIA. MINISTRY OF EDUCATION
Universalization and renewal of primary education in the perspective of an appropriate introduction to science and technology. Addis Ababa, Ministry of Education, 1984. 12 l., table. Reply to questionnaire
ED/BIE/CONFINTED/39/Q.1/83. (eng)
Microfiche: SIRE/01713

FINLAND. MINISTRY OF EDUCATION
Universalization and renewal of primary education in the perspective of an appropriate introduction to science and technology. Helsinki, Ministry of Education, 1983. 10, 4 p., table. Reply to questionnaire
ED/BIE/CONFINTED/39/Q.1/83. (eng)
Microfiche: SIRE/01716

FRANCE. NATIONAL COMMISSION FOR UNESCO
Généralisation et rénovation de l'enseignement primaire dans la perspective d'une initiation scientifique et technique appropriée. Paris, Commission française pour l'Unesco, 1983. 40 l. Reply to questionnaire
ED/BIE/CONFINTED/39/Q.1/83. (fre)
Microfiche: SIRE/01719

GABON. MINISTÈRE DE L'ÉDUCATION NATIONALE. DIRECTION GÉNÉRALE DES ENSEIGNEMENTS ET DE LA PÉDAGOGIE
Généralisation et rénovation de l'enseignement primaire dans la perspective d'une initiation scientifique et technique appropriée. Libreville, Direction générale des enseignements et de la pédagogie, Ministère de l'éducation nationale, Institut pédagogique national, 1984. 23 l. Reply to questionnaire
ED/BIE/CONFINTED/39/Q.1/83. (fre)
Microfiche: SIRE/01722

References

GERMAN DR. NATIONAL COMMISSION FOR UNESCO
Universalization and renewal of primary education in the perspective of an appropriate introduction to science and technology. Berlin, Commission of the German Democratic Republic for Unesco, 1983. 78 l., tables. Reply to questionnaire
ED/BIE/CONFINTED/39/Q.1/83. (eng)
Microfiche: SIRE/01724

GERMANY FR. STANDING CONFERENCE OF MINISTERS OF EDUCATION AND CULTURAL AFFAIRS OF THE LÄNDER IN THE GERMANY FR. SECRETARIAT
Universalization and renewal of primary education in the perspective of an appropriate introduction to science and technology/Universalisierung und Erneuerung der Primarerziehung im Hinblick auf eine angemessene Einführung in die Naturwissenschaften und die Technologie. Bonn, Secretariat of the Standing Conference of the Ministers of Education and Cultural Affairs of the Laender in the Federal Republic of Germany, 1983. 46 p., table. Reply to questionnaire
ED/BIE/CONFINTED/39/Q.1/83. (same text in eng, ger)
Microfiche: SIRE/01727

GUINEA. INSTITUT PÉDAGOGIQUE NATIONAL (GUINEA)
(Généralisation et rénovation de l'enseignement primaire dans la perspective d'une initiation scientifique et technique appropriée). Conakry, Institut pédagogique national, Commission nationale guinéenne pour l'Unesco, 1983. 6 l. Reply to questionnaire
ED/BIE/CONFINTED/39/Q.1/83. (fre)
Microfiche: SIRE/01731

GUYANA. MINISTRY OF EDUCATION, SOCIAL DEVELOPMENT AND CULTURE
Universalization and renewal of primary education in the perspective of an appropriate introduction to science and technology. Georgetown, Ministry of Education, Social Development and Culture, 1984. 16 l., fig., tables. Reply to questionnaire
ED/BIE/CONFINTED/39/Q.1/83. (eng)
Microfiche: SIRE/01734

HUNGARY. NATIONAL COMMISSION FOR UNESCO
Universalization and renewal of primary education in the perspective of an appropriate introduction to science and technology. Budapest, Hungarian National Commission for Unesco, 1983. ii, 10 l., table. Reply to questionnaire
ED/BIE/CONFINTED/39/Q.1/83. (eng)
Microfiche: SIRE/01737

INDIA. NATIONAL COMMISSION FOR UNESCO
Universalization and renewal of primary education in the perspective of an appropriate introduction to science and technology. New Delhi, Indian National Commission for Cooperation with Unesco, 1984. 23 l. Reply to questionnaire
ED/BIE/CONFINTED/39/Q.1/83. (eng)
Microfiche: SIRE/01740

INDONESIA. NATIONAL COMMISSION FOR UNESCO
The Universalization of primary education. Jakarta, Indonesian National Commission for Unesco, 1984. 44 l., figs., tables. Reply to questionnaire
ED/BIE/CONFINTED/39/Q.1/83. (eng)
Microfiche: SIRE/01743

IRAN (ISLAMIC REPUBLIC). NATIONAL COMMISSION FOR UNESCO
(Universalization and renewal of primary education in the perspective of an appropriate introduction to science and technology). Tehran, Iranian National Commission for Unesco, 1984. 9 l., table. Reply to questionnaire
ED/BIE/CONFINTED/39/Q.1/83. (eng)
Microfiche: SIRE/01746

IRAQ. MINISTRY OF EDUCATION. DIRECTORATE GENERAL OF CULTURAL RELATIONS. DEPARTMENT OF INTERNATIONAL AND ARAB ORGANIZATIONS
(Universalization and renewal of primary education in the perspective of an appropriate introduction to science and technology). Baghdad, Department of International and Arab Organizations, Directorate General of Cultural Relations, Ministry of Education, 1984. 26 l., tables. Reply to questionnaire
ED/BIE/CONFINTED/39/Q.1/83. (eng; also in ara)
Microfiche: SIRE/01749

IRELAND. DEPARTMENT OF EDUCATION
(Universalization and renewal of primary education in the perspective of an appropriate introduction to science and technology). Dublin, Department of Education, 1983. 8 l., fig., table. Reply to questionnaire
ED/BIE/CONFINTED/39/Q.1/83. (eng)
Microfiche: SIRE/01752

JAMAICA. MINISTRY OF EDUCATION
Universalization and renewal of primary education in the perspective of an appropriate introduction to science and technology. Kingston, Ministry of Education, 1983. 15 l., tables. Reply to questionnaire
ED/BIE/CONFINTED/39/Q.1/83. (eng)
Microfiche: SIRE/01757

JAPAN. NATIONAL COMMISSION FOR UNESCO
Universalization and renewal of primary education in the perspective of an appropriate introduction to science and technology. Tokyo, Japanese National Commission for Unesco, 1984. 22 l., tables. Reply to questionnaire
ED/BIE/CONFINTED/39/Q.1/83. (eng)
Microfiche: SIRE/01760

JORDAN. MINISTRY OF EDUCATION
Universalization and renewal of primary education in the perspective of an appropriate introduction to science and technology. Amman, Ministry of Education, 1983. 11 l. Reply to questionnaire
ED/BIE/CONFINTED/39/Q.1/83. (eng; also in ara)
Microfiche: SIRE/01763

KENYA. MINISTRY OF EDUCATION, SCIENCE AND TECHNOLOGY
Universalisation and renewal of primary education in the perspective of an appropriate introduction to science and technology. Nairobi, Ministry of Education, Science and Technology, 1984. 21 l., tables. Reply to questionnaire
ED/BIE/CONFINTED/39/Q.1/83. (eng)
Microfiche: SIRE/01766

KOREA R. KOREAN EDUCATIONAL DEVELOPMENT INSTITUTE (KOREA R)
Universalization and renewal of primary education in the perspective of an appropriate introduction to science and technology. Seoul, Korean Educational Development Institute, 1984. 69 p., table. Reply to questionnaire
ED/BIE/CONFINTED/39/Q.1/83. (eng)
Microfiche: SIRE/01769

KUWAIT. MINISTRY OF EDUCATION. DEPARTMENT OF PLANNIŃG
(Ta'mīm at-ta'līm al-ibtidā'ī wa-tağdīduhu bi-hadaf taḍmīnihi qadran mulā'iman min mabādi' al-'ilm wa-ät-tiknūlūǧiyya). al-Kuwayt, Idārat at-tahṭīṭ, Wazārat at-tarbiya, 1983. 10 l. Reply to questionnaire
ED/BIE/CONFINTED/39/Q.1/83. (ara)
Microfiche: SIRE/01772

LUXEMBOURG. MINISTÈRE DE L'ÉDUCATION NATIONALE
(Généralisation et rénovation de

References

l'enseignement primaire dans la perspective d'une initiation scientifique et technique appropriée). Luxembourg, Ministère de l'éducation nationale, 1984. 13 l. Reply to questionnaire ED/BIE/CONFINTED/39/Q.1/83. (fre)
Microfiche: SIRE/01775

MADAGASCAR. NATIONAL COMMISSION FOR UNESCO
Généralisation et rénovation de l'enseignement primaire dans la perspective d'une initiation scientifique et technique appropriée. Tananarive, Commission nationale malagasy pour l'Unesco, 1983. 13 l., tables. Reply to questionnaire ED/BIE/CONFINTED/39/Q.1/83. (fre)
Microfiche: SIRE/01776

MALAWI. NATIONAL COMMISSION FOR UNESCO
(Universalization and renewal of primary education in the perspective of an appropriate introduction to science and technology). Lilongwe, Malawi National Commission for Unesco, 1984. 9 l., tables. Reply to questionnaire ED/BIE/CONFINTED/39/Q.1/83. (eng)
Microfiche: SIRE/01778

MALAYSIA. NATIONAL COMMISSION FOR UNESCO
(Universalization and renewal of primary education in the perspective of an appropriate introduction to science and technology). Kuala Lumpur, Malaysian National Commission for Unesco, 1984. 46 p., diagrs., tables. Bibl.: p. 45–46. Reply to questionnaire ED/BIE/CONFINTED/39/Q.1/83. (eng)
Microfiche: SIRE/01781

MALTA. NATIONAL COMMISSION FOR UNESCO
(Universalization and renewal of primary education in the perspective of an appropriate introduction to science and technology). Valletta, Maltese National Commission for Unesco, 1983. 5 l. Reply to questionnaire ED/BIE/CONFINTED/39/Q.1/83. (eng)

Microfiche: SIRE/01784

MAURITIUS. NATIONAL COMMISSION FOR UNESCO
Universalization and renewal of primary education in the perspective of an appropriate introduction to science and technology. Port Louis, Mauritius National Commission for Unesco, 1983. 12 l. Reply to questionnaire ED/BIE/CONFINTED/39/Q.1/83. (eng)
Microfiche: SIRE/01787

MEXICO. SECRETARÍA DE EDUCACIÓN PÚBLICA. SUBSECRETARÍA DE EDUCACIÓN ELEMENTAL. DIRECCIÓN GENERAL DE EDUCACIÓN PRIMARIA
Generalización y renovación de la enseñanza primaria en la perspectiva de una iniciación científica y técnica apropiada. México, Dirección General de Educación Primaria, Subsecretaría de Educación Elemental, Secretaría de Educación Pública, Dirección General de Materiales Didácticos y Culturales, Secretaría de Educación Pública, Dirección General de Relaciones Internacionales, Secretaría de Educación Pública, 1983. 102 l., figs., tables. Reply to questionnaire ED/BIE/CONFINTED/39/Q.1/83. (spa)
Microfiche: SIRE/01790

MOROCCO. MINISTÈRE DE L'ÉDUCATION NATIONALE. DIRECTION DE LA PLANIFICATION. DIVISION DES ÉTUDES ET DU PLAN
Généralisation et rénovation de l'enseignement primaire dans la perspective d'une initiation scientifique et technique appropriée. Rabat, Division des études et du Plan, Direction de la planification, Ministère de l'éducation nationale, 1984. 12 l. Reply to questionnaire ED/BIE/CONFINTED/39/Q.1/83. (fre)
Microfiche: SIRE/01794

MOZAMBIQUE. MINISTÉRIO DA EDUCAÇÃO
(Universalization and renewal of primary education in the perspective of an appropriate introduction to science and technology). Maputo, Ministério da Educação, 1984. 30 l., tables. Reply to questionnaire ED/BIE/CONFINTED/39/Q.1/83. (eng)
Microfiche: SIRE/01797

NEPAL. MINISTRY OF EDUCATION AND CULTURE
Universalization and renewal of primary education in the perspective of an appropriate introduction to science and technology. Kathmandu, Ministry of Education and Culture, 1983. 5 l. Reply to questionnaire ED/BIE/CONFINTED/39/Q.1/83. (eng)
Microfiche: SIRE/01800

NETHERLANDS. MINISTRY OF EDUCATION AND SCIENCE. CENTRAL DIRECTORATE OF INTERNATIONAL RELATIONS. MULTILATERAL RELATIONS DIVISION
(Universalization and renewal of primary education in the perspective of an appropriate introduction to science and technology). The Hague, Multilateral Relations Division, Central Directorate of International Relations, Ministry of Education and Science, 1984. 18 l. Reply to questionnaire ED/BIE/CONFINTED/39/Q.1/83. (eng)
Microfiche: SIRE/01803

NEW ZEALAND. DEPARTMENT OF EDUCATION
Universalization and renewal of primary education in the perspective of an appropriate introduction to science and technology. Wellington, Department of Education, 1984. 21 l., tables. Reply to questionnaire ED/BIE/CONFINTED/39/Q.1/83. (eng)
Microfiche: SIRE/01805

NICARAGUA. MINISTERIO DE EDUCACIÓN
Generalización y renovación de la enseñanza primaria en la perspectiva de una iniciación científica y técnica apropiada. Managua, Ministerio de Educación, 1983. 36 l., table. Reply to questionnaire ED/BIE/CONFINTED/39/Q.1/83. (spa)
Microfiche: SIRE/01808

NIGERIA. NATIONAL COMMISSION FOR UNESCO
Universalization and renewal of primary education in the perspective of an appropriate introduction to science and technology. Lagos, Nigerian National Commission for Unesco, 1984. 26 l. Reply to questionnaire ED/BIE/CONFINTED/39/Q.1/83. (eng)
Microfiche: SIRE/01812

OMAN. NATIONAL COMMISSION FOR UNESCO
(Ta'mīm at-ta'līm al-ibtidā'ī wa-tağdīduhu bi-hadaf taḍmīnihi qadran mulā'iman min mabādi' al-'ilm wa-āt-tiknūlūğiyya). Masqaṭ, al-Laǧna al-waṭaniyya al-'umāniyya li-āt-tarbiya wa-āt-taqāfa wa-āl-'ulūm, 1984. ii, 14 l., tables. Reply to questionnaire ED/BIE/CONFINTED/39/Q.1/83. (ara)
Microfiche: SIRE/01816

PAKISTAN. NATIONAL COMMISSION FOR UNESCO
Universalization and renewal of primary education in the perspective of an appropriate introduction to science and technology. Islamabad, Pakistan National Commission for Unesco, 1984. 23 l. Reply to questionnaire ED/BIE/CONFINTED/39/Q.1/83. (eng)
Microfiche: SIRE/01819

PANAMA. MINISTERIO DE EDUCACIÓN
Generalización y renovación de la enseñanza primaria en la perspectiva de una iniciación científica y técnica apropiada. Panamá, Ministerio de Educación, 1984. 82 l., tables. Reply to

questionnaire
ED/BIE/CONFINTED/39/Q.1/83. (spa)
Microfiche: SIRE/01821

PARAGUAY. MINISTERIO DE EDUCACIÓN Y CULTO. DEPARTAMENTO DE ENSEÑANZA PRIMARIA
Generalización y renovación de la enseñanza primaria en la perspectiva de una iniciación científica y técnica adecuada. Asunción, Departamento de Enseñanza Primaria, Ministerio de Educación y Culto, 1983. 70 l, 13 p., tables. Reply to questionnaire
ED/BIE/CONFINTED/39/Q.1/83. (spa)
Microfiche: SIRE/01824

PERU. MINISTERIO DE EDUCACIÓN. DIRECCIÓN DE EDUCACIÓN PRIMARIA
Generalización y renovación de la enseñanza primaria en la perspectiva de la iniciación científica y técnica apropiadas. Lima, Direción de Educación Primaria, Ministerio de Educación, 1983. 20 l., table. Reply to questionnaire
ED/BIE/CONFINTED/39/Q.1/83. (spa)
Microfiche: SIRE/01826

POLAND. MINISTRY OF EDUCATION
(Généralisation et rénovation de l'enseignement primaire dans la perspective d'une initiation scientifique et technique appropriée). Varsovie, Ministry of Education, Department of the Organization of Research and Pedagogical Information, Institute of Pedagogical Studies, 1983. 33 l., table. Reply to questionnaire
ED/BIE/CONFINTED/39/Q.1/83. (fre)
Microfiche: SIRE/01829

PORTUGAL. MINISTRY OF NATIONAL EDUCATION
Généralisation et rénovation de l'enseignement primaire dans la perspective d'une initiation scientifique et technique appropriée. Lisbonne, Ministère de l'éducation, 1984. 35 l., tables. Reply to questionnaire
ED/BIE/CONFINTED/39/Q.1/83. (fre)
Microfiche: SIRE/01832

QATAR. MINISTRY OF EDUCATION
Universalization and renewal of primary education in the perspective of an appropriate introduction to science and technology. Doha, Qatar National Commission for Education, Culture and Science, 1984. 9 l., tables. Prepared by the Ministry of Education. Reply to questionnaire
ED/BIE/CONFINTED/39/Q.1/83. (eng; also in ara)
Microfiche: SIRE/01835

RWANDA. MINISTÈRE DE L'ENSEIGNEMENT SUPÉRIEUR ET DE LA RECHERCHE SCIENTIFIQUE
(Généralisation et rénovation de l'enseignement primaire dans la perspective d'une initiation scientifique et technique appropriée). Kigali, Ministère de l'enseignement supérieur et de la recherche scientifique, 1983. 17 l., tables. Reply to questionnaire
ED/BIE/CONFINTED/39/Q.1/83. (fre)
Microfiche: SIRE/01839

SAN MARINO. NATIONAL COMMISSION FOR UNESCO
(Généralisation et rénovation de l'enseignement primaire dans la perspective d'une initiation scientifique et technique appropriée). San Marino, Commissione nazionale sammarinese per l'Unesco, 1983. 8 l. Reply to questionnaire
ED/BIE/CONFINTED/39/Q.1/83. (fre)
Microfiche: SIRE/01841

SENEGAL. MINISTÈRE DE L'ÉDUCATION NATIONALE
(Généralisation et rénovation de l'enseignement primaire dans la perspective d'une initiation scientifique et technique appropriée). Dakar, Ministère de l'éducation nationale, 1984. 9 l. Reply to questionnaire
ED/BIE/CONFINTED/39/Q.1/83. (fre)
Microfiche: SIRE/01845

SEYCHELLES. MINISTRY OF EDUCATION AND INFORMATION
(Universalization and renewal of primary education in the perspective of an appropriate introduction to science and technology). Victoria, Ministry of Education and Information, 1983. 7 p., tables. Reply to questionnaire
ED/BIE/CONFINTED/39/Q.1/83. (eng)
Microfiche: SIRE/01848

SPAIN. MINISTERIO DE EDUCACIÓN Y CIENCIA. DIRECCIÓN GENERAL DE EDUCACIÓN BÁSICA
(Generalización y renovación de la enseñanza primaria en la perspectiva de una iniciación científica y técnica apropiada). Madrid, Dirección General de Educación Básica, Ministerio de Educación y Ciencia, 1983. 7 l. Reply to questionnaire
ED/BIE/CONFINTED/39/Q.1/83. (spa)
Microfiche: SIRE/01851

SRI LANKA. MINISTRY OF EDUCATION
Universalization and renewing of primary education in the perspective of an appropriate introduction to science and technology. Colombo, Ministry of Education, 1983. 26 l., tables. Reply to questionnaire
ED/BIE/CONFINTED/39/Q.1/83. (eng)
Microfiche: SIRE/01853

SUDAN. MINISTRY OF EDUCATION AND GUIDANCE. STRATEGY AND PLANNING ADMINISTRATION
Ta'mīm at-ta'līm al-ibtidā'ī wa-tağdīduhu bi-hadaf taḍmīnihi qadran mulā'iman min mabādi' al-'ilm wa-ăt-tiknūlūğiyya. al-Harṭūm, al-Idāra al-'āmma li-ăt-tahṭīṭ wa-ăl-istrātīğiyya, Wazārat at-tarbiya wa-ăt-tawğīh, 1984. ii, 13 p. Reply to questionnaire
ED/BIE/CONFINTED/39/Q.1/83. (ara)
Microfiche: SIRE/01856

SWEDEN. NATIONAL COMMISSION FOR UNESCO
(Universalization and renewal of primary education in the perspective of an appropriate introduction to science and technology). Stockholm, Swedish National Commission for Unesco, 1983. 6 l. Reply to questionnaire
ED/BIE/CONFINTED/39/Q.1/83. (eng)
Microfiche: SIRE/01859

SWITZERLAND. CONFÉRENCE SUISSE DES DIRECTEURS CANTONAUX DE L'INSTRUCTION PUBLIQUE
Blanc, Emile. *Généralisation et rénovation de l'enseignement primaire dans la perspective d'une initiation scientifique et technique appropriée.* Berne, Conférence suisse des directeurs cantonaux de l'instruction publique, 1984. 16 l. Reply to questionnaire
ED/BIE/CONFINTED/39/Q.1/83. (fre)
Microfiche: SIRE/01861

SYRIAN AR. MINISTÈRE DE L'ÉDUCATION
(Généralisation et rénovation de l'enseignement primaire dans la perspective d'une initiation scientifique et technique appropriée). Damas, Ministère de l'éducation, 1984. 14 l., table. Reply to questionnaire
ED/BIE/CONFINTED/39/Q.1/83. (fre; also in ara)
Microfiche: SIRE/01864

TANZANIA UR. MINISTRY OF NATIONAL EDUCATION
Universalization and renewal of primary education in the perspective of an appropriate introduction to science and technology. Dar es Salaam, Ministry of National Education, 1984. 11 l., tables. Reply to questionnaire
ED/BIE/CONFINTED/39/Q.1/83. (eng)
Microfiche: SIRE/01867

THAILAND. NATIONAL COMMISSION FOR UNESCO
Universalization and renewal of primary education in the perspective of an appropriate introduction to science and technology. Bangkok, Thailand National Commission for Unesco, 1983.

References

21 l., tables. Reply to questionnaire
ED/BIE/CONFINTED/39/Q.1/83. (eng)
Microfiche: SIRE/01870

TOGO. MINISTÈRE DE L'ÉDUCATION NATIONALE ET DE LA RECHERCHE SCIENTIFIQUE
Généralisation et rénovation de l'enseignement primaire dans la perspective d'une initiation scientifique et technique appropriée. Lomé, Ministère de l'éducation nationale et de la recherche scientifique, 1984. 4 l., tables. Reply to questionnaire
ED/BIE/CONFINTED/39/Q.1/83. (fre)
Microfiche: SIRE/01872

TONGA. MINISTRY OF EDUCATION
(Universalization and renewal of primary education in the perspective of an appropriate introduction to science and technology). Nuku'alofa, Ministry of Education, 1983. 5, 9 l., tables. Reply to questionnaire
ED/BIE/CONFINTED/39/Q.1/83. (eng)
Microfiche: SIRE/01874

TUNISIA. MINISTÈRE DE L'ÉDUCATION NATIONALE
(Généralisation et rénovation de l'enseignement primaire dans la perspective d'une initiation scientifique et technique appropriée). Tunis, Ministère de l'éducation nationale, 1983. 10, 3 l., tables. Reply to questionnaire
ED/BIE/CONFINTED/39/Q.1/83. (fre)
Microfiche: SIRE/01877

TURKEY. NATIONAL COMMISSION FOR UNESCO
La Généralisation et la rénovation de l'enseignement primaire dans la perspective d'une initiation scientifique et technique appropriée. Ankara, Commission nationale turque pour l'Unesco, 1984. 52 l., tables. Reply to questionnaire
ED/BIE/CONFINTED/39/Q.1/83. (fre)
Microfiche: SIRE/01880

UGANDA. MINISTRY OF EDUCATION
Universalization and renewal of primary education in the perspective of an appropriate introduction to science and technology. Kampala, Ministry of Education, 1983. 25 l., table. Reply to questionnaire
ED/BIE/CONFINTED/39/Q.1/83. (eng)
Microfiche: SIRE/01883

UK. OVERSEAS DEVELOPMENT ADMINISTRATION
Universalization and renewal of primary education in the perspective of an appropriate introduction to science and technology. London, Overseas Development Administration, 1984. 13, iii l. Reply to questionnaire
ED/BIE/CONFINTED/39/Q.1/83. (eng)
Microfiche: SIRE/01886

UKRAINIAN SSR. MINISTERSTVO PROSVEŠČENIJA
Vseobščee rasprostranenie i obnovlenie načal'nogo obrazovanija putem vvedenija sootvetstvujuščih elementov naučnyh i tehničeskih znanij. Kiev, Ministerstvo prosveščenija, 1983. 17 l., table. Reply to questionnaire
ED/BIE/CONFINTED/39/Q.1/83. (rus)
Microfiche: SIRE/01889

UNITED ARAB EMIRATES. MINISTRY OF EDUCATION. GENERAL DIRECTORATE FOR EDUCATIONAL PLANNING
Ta'mīm at-ta'līm al-ibtidā'ī wa-tağdīduhu bi-hadaf taḍmīnihi qadran mulā'iman min mabādi' al-'ilm wa-ăt-tiknūlūğiyya. Abū Ẓabī, al-Idāra al-'āmma li-ăt-tahṭīṭ at-tarbawī, Wazārat at-tarbiya wa-ăt-ta'līm, 1983. ii, 14 l., table. Reply to questionnaire
ED/BIE/CONFINTED/39/Q.1/83. (ara)
Microfiche: SIRE/01892

USA. DEPARTMENT OF EDUCATION
Universalization and renewal of primary education in the pespective of an appropriate introduction to science

and technology. Washington, DC, United States Department of Education, 1984. 30 l. Reply to questionnaire
ED/BIE/CONFINTED/39/Q.1/83. (eng)
Microfiche: SIRE/01895

USSR. MINISTERSTVO PROSVEŠČENIJA
Vseobščee rasprostranenie i obnovlenie načal'nogo obrazovanija putem vvedenija sootvetstvujuščih elementov naučnyh i tehničeskih znanij. Moskva, Ministerstvo prosveščenija, 1983. 13 l., table. Reply to questionnaire
ED/BIE/CONFINTED/39/Q.1/83. (rus)
Microfiche: SIRE/01898

VENEZUELA. MINISTERIO DE EDUCACIÓN
(Generalización y renovación de la enseñanza primaria en la perspectiva de una iniciación científica y técnica apropiada). Caracas, Ministerio de Educación, 1983. 9 l. Reply to questionnaire
ED/BIE/CONFINTED/39/Q.1/83. (spa)
Microfiche: SIRE/01901

VIET NAM SR. NATIONAL COMMISSION FOR UNESCO
Answers to questions on universalization and innovation of primary education in the prospect of suitable scientific and technical initiative. Hanoi, Vietnamese National Commission for Unesco, 1984. 8 l., table. Reply to questionnaire
ED/BIE/CONFINTED/39/Q.1/83. (eng)
Microfiche: SIRE/01903

YUGOSLAVIA. NATIONAL COMMISSION FOR UNESCO
Šoljan, Nikša Nikola. *Universalization and renewal of the primary education in the perspective of an appropriate introduction to science and technology.* Zagreb, Republic Institute for Education of the SR of Croatia; Belgrade, Yugoslav Commission for Unesco, 1984. 20 p. Reply to questionnaire
ED/BIE/CONFINTED/39/Q.1/83. (eng)

Microfiche: SIRE/01907

ZAMBIA. NATIONAL COMMISSION FOR UNESCO
(Universalization and renewal of primary education in the perspective of an appropriate introduction to science and technology). Lusaka, Zambia National Commission for Unesco, 1983. 8 l. Reply to questionnaire
ED/BIE/CONFINTED/39/Q.1/83. (eng)
Microfiche: SIRE/01911

ZIMBABWE. MINISTRY OF EDUCATION AND CULTURE
(Universalization and renewal of primary education in the perspective of an appropriate introduction to science and technology). Harare, Ministry of Education and Culture, 1984. 6 l. Reply to questionnaire
ED/BIE/CONFINTED/39/Q.1/83. (eng)
Microfiche: SIRE/01914